Agriculture and Horticulture in New Zealand

Agriculture and Horticulture in New Zealand

EDITED BY KEVIN STAFFORD

MASSEY UNIVERSITY PRESS

Contents

Foreword

We all enjoy eating high-quality food, wearing natural fibres, using timber products, seeing flowers on display and living in landscaped surroundings. This book will show you how agricultural and horticultural produce is grown, managed and harvested so that it can be sent around New Zealand and the world to provide naturally produced, premium-quality food and products that make life more enjoyable.

Most agricultural and horticultural production in New Zealand is exported and so must meet stringent health, biosecurity, animal welfare and environmental regulations. A useful way to understand the land-based systems used to produce food and other products is to work backwards from what the customer and regulators require, to how to grow and manage crops, trees and livestock that meet those requirements. This value chain, from customers to marketers to transport logistics to processing and packaging businesses to farmers and orchardists, is what New Zealand is a world leader in, and a key to our prosperity as a nation.

Over the approximately 800 years that people have been growing crops in New Zealand we have been learning how to get the best from our soil and climate to consistently grow high-yield crops and pastures in a sustainable way. Mistakes have been made along the way, but the information in this book will set you on the path to understanding the great challenge of sustainably using our natural environment while feeding people with fresh, healthy and safe food.

Most agricultural and horticultural production is land-based, although there is increased interest in indoor farming worldwide. While New Zealand is mainly a land-based producer, our use of land is dynamic — responding to market signals, new technologies and climate change. New Zealand has a total land area of 26.8 million hectares (ha), and approximately half is used for primary industry production, with approximately 45 per cent used for agricultural or horticultural production and 6 per cent for planted forestry. Livestock grazing on pasture and forage crops is our largest land use, with 92 per cent of the agricultural and horticultural land (11.1 million ha) growing pasture and forages. The major use of pastoral land is for sheep and beef production, but the area used for this has been declining over the past 20 years or so; until recently, there was strong growth in the area of land used for dairying. However, although the area of land used for horticulture and arable cropping is relatively small, it has been increasing and is predicted to continue to increase.

The area of land used for different agricultural and horticultural enterprises only tells part of the story. Productivity per hectare has greatly increased in all agricultural and horticultural enterprises over the decades and continues to do so. In the case of apples, kiwifruit and grain crops, both production and productivity have been increasing due to plant breeding, use of technologies such as irrigation, sensors and automation, and improved plant management based on scientific research.

Advances in apple production are a good example of research and innovation resulting in improvements along the value chain. Yields of apples have dramatically increased due to high-density plantings of apple trees trellised on wires so that the plants intercept more light. The apples produced on the rows of trees are more accessible for semi-automated picking, and ultimately full robotic picking. The apple varieties are bred to provide apples that different markets prefer, and they grow on trees kept short by the type of rootstock they are grafted on to. Pest and disease occurrences are predicted using software programs in computers connected to sensors in the orchard, and spraying is minimised through the use of integrated pest management. Picked apples are sorted by size, colour and blemishes using arrays of high-speed cameras feeding data into computer programs, and are then automati-

cally packed into boxes the right size for each apple category. The cardboard boxes are designed to prevent bruising and to keep the apples fresh as they are transported around the world.

New Zealand's relatively small area of arable and horticultural crops produces world-leading yields per hectare, for example, in crops such as kiwifruit, apples, maize, wheat and barley. The total horticultural production and the economic value of those products has been on the rise over the past 20 years and there is much optimism that our horticultural export earnings will continue to rise due to high overseas demand for our high-quality and superbly marketed fruit, vegetables, seeds, grain, flowers and wine.

Given that agriculture and horticulture are major drivers of the New Zealand economy and are such dynamic and innovative industries, albeit with some environmental challenges, it is not surprising that the industry supports careers across the spectrum of growers, managers, agricultural and environmental consultants, agricultural service providers, engineers, agribusiness providers, and marketers. All these careers contribute to the challenge of feeding the people of the world high-quality food while sustaining our environment.

As you read through this book and marvel at the innovation and ingenuity behind agriculture and horticulture in New Zealand, consider how these industries contribute to the quality of your life.

Peter Kemp
Professor of Pasture Science
School of Agriculture and Environoment
Massey University

Chapter 1
Soils

Alan Palmer &
Dave Horne

Chapter 1
Soils

Alan Palmer and Dave Horne

School of Agriculture and Environment, Massey University

Introduction

Soils are a product of the environment in which they form — an environment that may evolve with time because of many factors, for example climate change. Soils, which vary from naturally occurring to heavily modified bodies, are where earth, climatic and biological processes intertwine. Our modern understanding of soil formation and classification has been influenced by two scientists: a Russian named Vasily Dokuchaiev, in the latter part of the nineteenth century; and an American, Hans Jenny, in the 1940s. Their thoughts have given us the famous equation:

$$S = f(cl, o, r, p, t)$$

This is called the 'soil-forming factors' or 'environment of soil formation' equation, where S = soil, cl = climate, o = flora and fauna, r = relief or topography, p = parent material and t = time available for formation of the soil. Some soil scientists now add an anthropogenic factor (a) to the list, while others accommodate the human impact on soils in the organisms (o) factor. What this means is that if we are blessed with a little knowledge of each of the soil-forming factors, we should be able to predict the type of soil present beneath our feet.

It quickly becomes apparent that specific soils occur on specific landscape units. We would expect to find different soils on a river plain that is subject to regular flooding than on a higher terrace where flooding no longer occurs. Similarly, in hill country a soil forming on a dry ridge is unlikely to resemble one forming in a moist swale at the foot of the slope. A knowledge of landscapes (geomorphology) is essential in understanding soil distribution.

We can learn a lot about New Zealand soils by considering each of the soil-forming factors

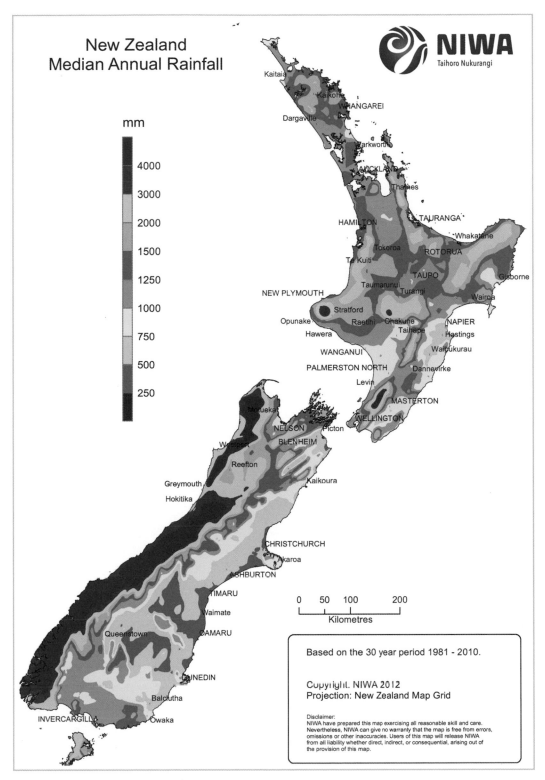

Figure 1.1 Median annual rainfall in New Zealand. (Source: Niwa)

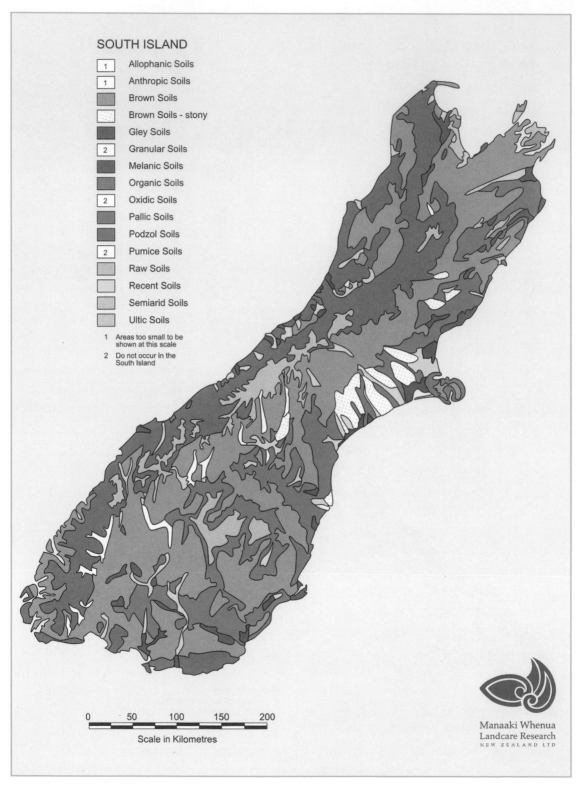

Figure 1.2 Soil map of the South Island. (Source: Manaaki Whenua Landcare Research)

in turn, while recognising that all of the factors interact to determine the type of soil present at any site in the landscape.

Climate

Due to its maritime setting in the middle latitudes and the dominant westerly flow of weather across the country, New Zealand has a temperate climate. Warm oceans surrounding the country ensure that incoming air is humid. The New Zealand landmass is also mountainous, particularly along its axial spine. Humid air is forced upwards by the elevated terrain, resulting in orographic rainfall (Figure 1.1). Nowhere is this more dramatically demonstrated than on the West Coast of the South Island. Moisture-laden winds coming in from the Tasman Sea drop 1500–3000 mm rainfall at the coast. Inland, 8000 to 14,000 mm per year can be recorded in the western foothills of the Southern Alps. It is interesting to note that the maximum rainfall is west of the highest parts of the Alps. A few tens of kilometres east of the main divide, the descending winds are dry and rainfall is commonly less than 1000 mm. The west coast of both islands is generally moister than the east coast, but local topography may superimpose an orographic effect on regional rainfall patterns.

The influence of rainfall on soil distribution is clearly demonstrated in the southern South Island (Figure 1.2). Central Otago is an area of basins and ranges in schist and greywacke–argillite bedrock. Being on the eastern side of the main divide, it is in a rain-shadow area and, on average, receives as little as 330 mm of precipitation per annum. Semiarid soils in New Zealand are soils formed on bedrock, slope materials, alluvium or loess, and where there has been sufficient time to weather the materials from which the soils form (soil parent materials) and where the annual rainfall is less than about 500 mm. Characteristically there is insufficient rainfall to carry all of the soluble weathering products, such as calcium carbonate, from the profile, resulting in whitish powdery coatings in the soil. Where annual rainfall is between approximately 500 and 1000 mm on the same parent material, pale-coloured soils called Pallic soils occur (Figure 1.3). Ironically, despite the rainfall still being low, these soils tend to have impeded drainage due to the way the soils are forming. Whereas Semiarid soils are seldom saturated, Pallic soils are often saturated in winter but dry in summer. At rainfalls higher than approximately 1000 mm on the same soil parent material, the soils stay moist for a longer time (due in part to the greater rainfall) but, counterintuitively, are often well drained.

This relationship between soils on the same parent material and in similar landscapes but under contrasting climate illustrates the importance of soil moisture in forming soils and developing soil properties.

Organisms

Organisms are flora and fauna that live on and within the soil. Pre-human New Zealand was unusual in being dominantly forested and having few soil-living animals other than birds and reptiles. Now a large proportion of our landscape is farmed, and humans and other animals living on the soil have modified it significantly. In pre-human New Zealand, the moister west

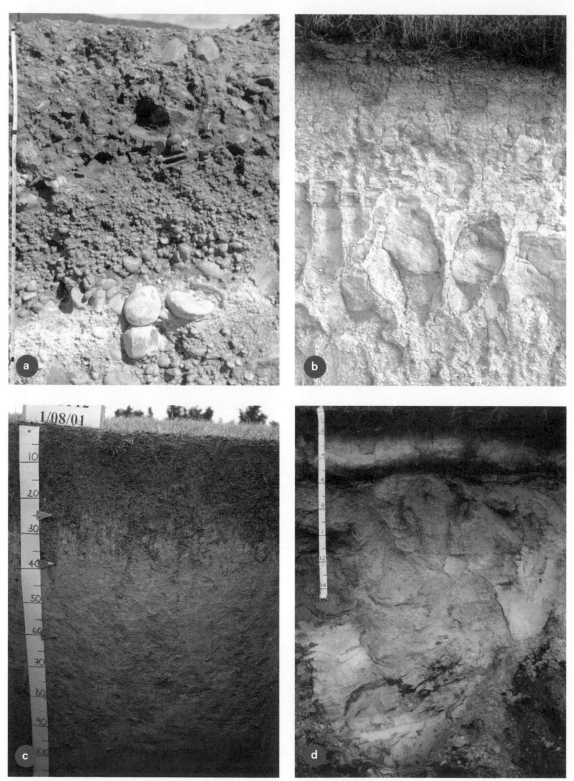

Figure 1.3 Southland soils: (a) semi-arid; (b) Pallic soil; (c) brown soil; (d) Podzol soil.

coast of both main islands, and much of the east, was cloaked with thick rainforest. Exceptions were the drier basins of Central Otago, Canterbury, Marlborough, Wairarapa and Hawke's Bay. In these areas, where rainfall was below approximately 1000 mm, the forest opened up progressively to more-scattered trees, shrubland and tussock-grassland. The vegetation present was closely related to rainfall and altitude, but landscape stability was also very important.

The type of forest that grew at any location has had a lasting legacy on the properties of some New Zealand soils. Areas once covered with podocarp (e.g. rimu, mataī, miro and kauri) and beech forests are often podzolised, resulting in Podzol soils. The word *podzol* has its roots in the Russian language, where it means ash. Villagers in the pine forests of northern Russia once collected pine needles for animal bedding. Underneath the thick layer of needles they would often uncover whitish material that they thought was ash from previous forest fires. We now know that pine trees, podocarps and other related families of vegetation produce certain chemicals in their leaves and bark. Rainwater carries these chemicals (among them poly-phenols) into the soil, where they cause metal cations to move into solution and then down the soil profile. Iron and aluminium are particularly affected and their absence from the soil results in a bleached, whitish colour. The soil is also acidic and leached of other nutrient cations (such as calcium, potassium and magnesium) and anions (such as phosphate and nitrate). The plant is essentially removing competitors by being adapted to recycle nutrients through feeder roots in the plant litter that it, itself, has shed.

Even a hundred years after clearance, the legacy of podzolisation under the previous

Figure 1.4 Mollisol soils (a) occur on natural grasslands such as this example from Russia. We do not have this type of natural grassland soils in New Zealand. Note that the topsoil is almost 1 metre thick. Most of the organic matter is in the soil. Podzol soils (b) occur in forests, for example, in this coniferous forest in Germany. Most of the organic matter is in the tree litter on the forest floor.

podocarp or beech forest cover can be seen in poorer soil fertility than in areas where other forest trees such as broadleaf species predominated (Figure 1.4). This concept is encapsulated by the following terms. *Mull* vegetation includes broadleaf tree species that dwell on more-fertile sites where leaf litter is broken down by an active soil flora and fauna, returning nutrients to the soil. Conversely, *Mor* vegetation favours and encourages less-fertile sites where acidity is higher, flora and fauna are less active, and leaf litter accumulates at the soil surface. These species then recycle nutrients from this leaf litter.

The process of podzolisation can be faster or slower depending on the substrate (soil parent material) that the soil has formed from and the climate. For example, podzolisation has occurred in less than 2000 years on Taupō Pumice in higher-rainfall areas under podocarp forest. Taupō Pumice is both coarse-textured and siliceous, which hastens the podzolisation process.

Around the world, soils that form beneath forests are fundamentally different to those that form under natural grasslands. In both cases, photosynthesis allows plants to grow and take in carbon dioxide from the atmosphere to form organic matter. In the case of a forest this organic matter is the wood, leaves and roots. In a forest ecosystem a much larger proportion of the organic matter is stored in the actual plant and in litter that is dropped to the forest floor and not yet fully incorporated into the soil. Relatively little is found in soil horizons. In a grassland ecosystem, a relatively small proportion of organic matter is held in the above-ground plant or its roots. Plant residues, and dung from the animals grazing the grasses and herbs, are quickly returned to the soil, where incorporation is assisted by soil biota. A much larger proportion of the ecosystem's carbon is stored within the soil rather than in the plant. Ecosystems, from tundra in the Arctic to various types of forest and grasslands or wetlands, vary considerably in their biomass and the rate of turnover of that biomass. This difference is clearly expressed in the soils that form.

Everywhere in the world where humans have changed or adapted the natural vegetation, the soils have also changed. The soils we farm are different to how they were in their natural state, and have considerably different biota. It is debatable whether human use of soils has improved or degraded them; both answers can be true for different locations and circumstances. On the one hand, addition of fertilisers, lime, organic residues, irrigation water and artificial drainage has made many soils more productive. Conversely, human use has also resulted in accelerated erosion, compaction, loss of organic matter in cropped soils, structural degradation, the addition of potentially harmful waste products, over-fertilisation, and use of pesticides, fungicides and insecticides. The biodiversity and natural functioning of farmed soils has been affected. An example is the hill and steep-land soils of the North Island of New Zealand. When European farmers first settled the thickly forested hill and steep-land areas from the late 1800s to early 1900s, they logged and burned the forest cover to produce grassland for their animals. The soils they encountered had developed in unison with the forest after the Last Glacial Period. During the Last Glacial, from about 30,000 to 15,000 years ago it was too cold for extensive forests to grow except in the northern part of the island. Much of the hill and steep-land at this time had a shrub or grassland cover, soils were thin and erosion rates were extreme. As the climate warmed at the end of the Last Glacial, the forest grew and the roots of the trees held the developing soil to the slopes. Erosion rates under a forest cover are as little as 10 per cent of those under grass cover, although the local influence of the type of geology, tectonics and climate are also important in determining the rate of natural erosion. Therefore, 20–30 years after clearance, as the roots of the former indigenous forest trees decayed, the binding force that was holding the soil to the slope was removed. Introduced pasture grasses have relatively shallow and weak roots that are unable to hold soils in place when those soils are saturated by winter rain or tropical summer storms. There is a threshold of between 120 and 200 mm rainfall, depending on slope, aspect and geology, above which soil-slip erosion occurs in a concerted and destructive manner. Cycles of soil-slip erosion, in some cases occurring every 10–20 years, have stripped the soil cover back to the underlying bedrock in places. Some slopes have lost as much as 60 per cent of the spatial soil cover that they had at forest clearance. Work by Landcare Research has shown that not only does the soil re-form very slowly, perhaps over hundreds of years, but pasture productivity is permanently affected too (Figure 1.5). This is because the re-forming soils are thinner and less able to store moisture for summer pasture growth; they also contain less carbon, have lost the fertiliser nutrients and lime applied over preceding years, and the soil biology that aids in nutrient cycling has been disrupted.

Relief

There is a strong relationship between the landscape and the soil that forms upon it. Pedologists mapping soils are the first to understand the landscape around them. The words *topography*

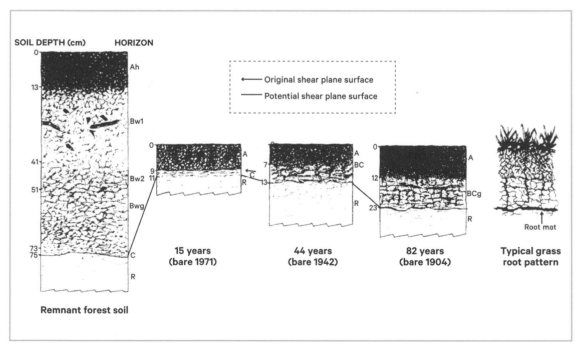

Figure 1.5 Erosion — mostly by soil slips — of steep-land soils removes the original soil that developed under forest. The soil reforms slowly and it takes thousands of years for it to fully reform. In many places the erosion of steep-land soils is unsustainable. A shear plane is a plane of likely failure when a slip occurs (the material above the shear plane slides off). Note: A is farmed topsoil; Ah is unfarmed topsoil; BC is weakly weathered; BCg is a weakly weathered horizon with mottles indicating impeded drainage Bw is well-drained brown subsoil; Bwg is a mottled horizon of imperfect drainage; C is unweathered slope material; R is rock (in this case mudstone). (After Trustrum and de Rose 1988)

or *relief* are used to describe the shape of the land. Individual pieces of land with common formation and relief are called *landforms*. The shape of a landform can have a strong influence on how surplus water from rainfall moves through or across the soil, where moisture is shed or percolates through the soil, where it is transmitted laterally, and where it accumulates. We usually see this manifest as soil drainage. For example, higher and convex parts of the landscape are often well drained, while concave and lower parts are more poorly drained.

Water in soil is a major determinant of how it forms. All soils retain a certain amount of rainfall within the soil (*storage*), allow some water to pass through laterally and horizontally (*drainage*), and have water leave the soil and move back into the atmosphere, mostly by transpiration through plant leaves. In most circumstances where plants cover the surface, relatively little water is directly evaporated from the soil surface. Transpiration and evaporation are collectively known as *evapotranspiration*. The proportion of storage, drainage and evapotranspiration varies greatly according to all of the soil-forming factors. Topography and soil texture (proportion of gravel, sand, silt or clay) strongly influence the drainage of the soil in any climatic regime.

Nowhere is the relationship between soil and topography better expressed than in areas of sand dunes, for example those on the south-west coast of the North Island, from Whanganui to Paekākāriki. Here, sand dunes have been accumulating since the Last Glacial Period, 30,000 years ago, and more particularly in the last 7000 years since the sea reached close to its current level after the glacial ice had melted. The sand is derived from soil and rock material eroded from the steep-lands and mountains of central North Island and transported to the coast by rivers, for example the Whanganui, Rangitīkei and Manawatū. The sand is then pushed south along the coast by wave action. For a variety of reasons the dune-building has waxed and waned in intensity and volume, resulting in phases of sand accumulation over time. (We will consider the age of the dunes in further discussion below.) The sandy soils of the dunes are porous and water passes through them readily, so much so that little is retained for plant growth and the dunes then appear droughty in dry summer months. However, the lower areas between the dunes, known as sand plains, are closer to the water-table and are often saturated, particularly in winter. These areas are low-lying, sometimes close to sea-level, and natural drainage pathways towards the coast can be blocked by younger dunes.

Even in low-rainfall areas, such as coastal Manawatū, the water-table can be close to the surface in wet seasons and when evapotranspiration is low in winter. Therefore, on any sand dune and adjacent sand plain, the soil that forms is dictated primarily by its position in the landscape (Figure 1.6). Soils on the upper parts of the dune drain freely, leach nutrients and are drought-prone in summer. Near the base of the dune the soils drain freely in summer, when the water-table is low, but come under the influence of a fluctuating water-table in their subsoil in winter. Soils on the sand flats have a fluctuating water-table that might rise close to the surface in winter. In particularly low-lying areas where the water-table remains close to the surface for most of the year, the soils are poorly drained, nutrients leach only slowly, and organic matter from plant residues in the topsoil do not oxidise and so may accumulate as peat.

Figure 1.6 A sequence of soils in 500–1800-year-old sand dunes in Manawatū: (a) the landscape in this area; (b) well-drained Motuiti soils that occur on the dune; (c) imperfectly drained Himatangi soils that occur on the higher parts of the sand plain between dunes; (d) Pukepuke soils — poorly drained soils on the sand plain that form where water-tables are high. This sequence of similar-aged soils with the same parent material but different properties is called a 'toposequence', in which the soils have formed due to their position in the landscape.

Another dramatic New Zealand landscape where the influence of topographic position on soil formation is clearly seen is in western Taranaki. Here, in the shadow of Mt Taranaki, there is an unusual hummocky landscape of 5–20 m high, 1000–10,000 m2 rounded hummocks with flatter land between. The hummocks are composed of chaotic piles of rock derived from the collapse of former versions of the strato-volcano Mt Taranaki. In its approximately 130,000-year life, Mt Taranaki has built up through eruption and subsequently catastrophically collapsed as many as 13 times, or once every 10,000 years or so. Indeed, the major volcanic hazard of Mt Taranaki is not necessarily eruption but the collapse through debris avalanche during its partial destruction.

The hummocky remains of older debris avalanches are now more rounded and the landscape is softened by the volcanic ash deposited in subsequent eruptions. North and east of the volcano the ash is now metres deep, but to the west where younger collapses have been directed, the ash is thinner or even absent so the bouldery remains are often visible. Each of the debris avalanche hummocks has a sequence of soils depending on position on or between the mounds, and on the thickness of ash cover. One such sequence of soils is found on hummocks from debris avalanches that occurred 16,000–20,000 years ago. Since the collapse, 60–100 cm of ash has accumulated on the land surface, so the soils are formed in this ash rather than the underlying bouldery material.

On the top and sides of the mounds, the soils are well drained, signified by rich brown colours in the subsoil. Near the base of the mounds, the water-table again fluctuates between summer and winter, leaving a mottled appearance of grey and orange colours in the subsoil. These mottles are caused by the changes in the status of iron in the ash soil parent material. The poorly drained soils between the mounds, where water-tables are high, are almost unrecognisable as being formed in the same ash soil parent material. The ash is now grey, with some orange mottles due to the influence of iron-altering processes in saturated soil below the water-table (Figure 1.7).

On every hillside, no matter where in the world, there is some degree of relationship between the soil that is forming and its position in the landscape. On our hills and steep-lands discussed earlier, there is often a negative relationship between soil thickness and development (and presence even) versus increasing slope.

Parent material

Parent material is what the soil has formed from. In some cases this is the bedrock below; in other cases it is material that has been transported to the site, for example deposits from a river. Rock types vary in their geochemical composition, mineralogy, hardness and texture. Their hardness and erodibility may also be influenced by physical, chemical and biological weathering and tectonic forces such as uplift, faulting and folding. The mechanism by which rocks erode — whether it be mass wasting under the influence of gravity, ice or liquid, water or wind — also determine the types of parent material that a soil forms from.

New Zealand has excellent examples of the three main kinds of rock: igneous, metamorphic and sedimentary. Igneous rocks are either crystalline rocks that have cooled from magma within

Figure 1.7 (a) Mt Taranaki and the debris avalanche mounds; (b) Warea; (c) Tipoka; (d) Awatuna soils.

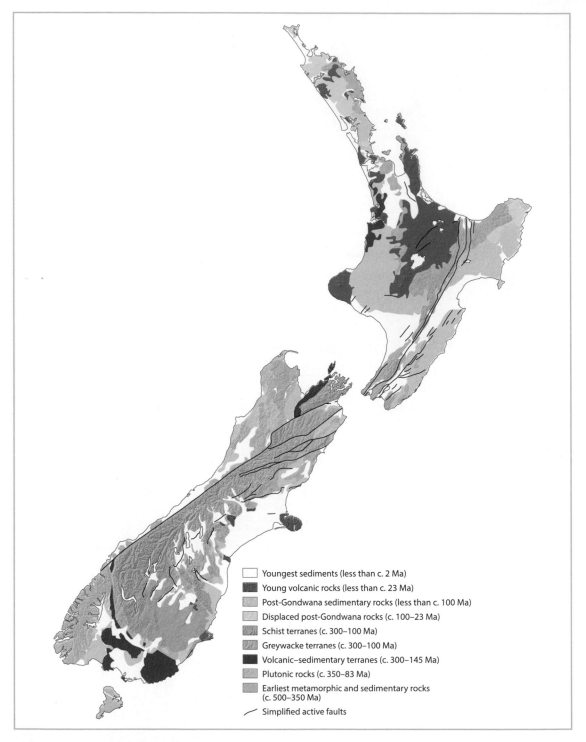

Figure 1.8 A simplified geology map of New Zealand. The rocks influence the soils that form, although soils can form on materials transported long distances by wind (volcanic ash and loess) or by water (alluvium). The oldest rocks are in Fiordland, the West Coast and Nelson. The axial ranges are schist (purple) and greywacke–argillite (blue). The young soft rock steep-land is orange and part of the yellow colour. Young volcanic rocks are shown in red. (Source: Isaac 2015)

the Earth's crust (plutonic) or volcanic rocks that have erupted at or near the surface. Plutonic igneous rocks are common in the north-west and south-west of the South Island, and in smaller areas of Northland in the North Island. Volcanic rocks dominate the central North Island, parts of Northland and eastern parts of the South Island.

Metamorphic rocks are either igneous or sedimentary rocks that have been subject to re-heating adjacent to an intrusion of magma, or deeply buried and subject to both heating and pressure. Given that both degree of heating and pressure from depth of burial vary, there are degrees of metamorphism from weak to strong. Metamorphic rocks are common in central and northern South Island, while weakly metamorphic rocks are found in places along the axial ranges of the North Island, resulting from deeply buried sedimentary rocks.

Sedimentary rocks are derived from eroded and fragmented igneous and metamorphic rocks. The main axial ranges of the North Island and most of the Southern Alps in the South Island are composed of two interbedded sedimentary rock types called greywacke and argillite, which were laid down 150–250 million years ago in an ocean basin adjacent to the ancient continent of Gondwana. Since this time, the greywacke and argillite rocks have twice been uplifted, faulted and folded along a continental plate boundary, and subsequently eroded, forming more sedimentary rocks of younger age.

The younger cycle of active plate-boundary formation and tectonic disruption began about 25 million years ago and is ongoing, with uplift and mountain-building particularly active in the past 5 million years. Sedimentary basins, which are collection areas for eroded sediments, formed around New Zealand, particularly adjacent to the emerging axial ranges. In places the remains of calcareous marine organisms, including molluscs, barnacles and bryozoa accumulated in preference to the sediment to form limestones. The once marine sedimentary basins themselves are now being uplifted above sea level. Once uplifted, the soft marine mudstones and sandstones, and somewhat harder limestones, are easily eroded by streams cutting down into them. This accounts for the soft-rock, easily erodible hill and steep-land found in many parts of New Zealand (Figure 1.8).

Sedimentary rocks are composed of fragments of the rocks from which they were sourced, minus a proportion of easily weatherable minerals that were removed or altered during erosion, transport or deposition. In this way, sedimentary rocks become dominated by less-weatherable minerals. Most of our greywacke–argillite rocks and younger mudstones and sandstones are composed primarily of minerals called quartz and feldspar. These minerals contain few of the nutrients that our farmed plants (grass, arable and horticultural crops) need to flourish, so nutrients need to be added to the soil for economic returns. Rocks and other soil parent materials such as alluvium that are dominated by quartz and feldspar are called *quartzo-feldspathic* and are dominant in New Zealand.

In the central North Island and a few locations elsewhere in New Zealand, volcanic rocks dominate. These include the white pumices erupted from Taupō and near Rotorua, the volcanic ash from strato-volcanoes such as Ruapehu and Taranaki that weather yellowish brown, and the localised scoria from basaltic volcanoes near Auckland and Northland which weather to a reddish-brown colour. The colour is important as it gives an indication of the relative amounts

of iron in the deposits. Volcanic parent materials also contain volcanic glass, a non-crystalline material that solidifies from rapidly cooled magma. When this glass weathers, it can form another non-crystalline clay-sized material called allophane. Allophane has the ability to adsorb anions, especially phosphate which is found naturally in the soil and also applied in some fertilisers. The adsorbed phosphorus is held tightly by the allophane and released in only small amounts for plant growth. In contrast, quartzo-feldspathic parent material contains crystalline clays that exchange cations such as $Ca2+$, $Mg2+$, $K+$ and $Na+$ with the water in soil, thus providing nutrients for plant growth. In most situations, crystalline clays do not adsorb or only weakly adsorb anions such as phosphate. Therefore, a dairy farmer on soils derived from volcanic materials in Taranaki, Waikato or Bay of Plenty may have to apply as much as twice the quantity of phosphate-based fertiliser as a dairy farmer on quartzo-feldspathic parent material in Manawatū, Canterbury or Southland.

Time

All soils take time to form. Soils can form from bedrock or deposited sediment such as alluvium from river flooding, or dune sand. The bedrock or sediment may be completely fresh, or it may be pre-weathered in which case soil formation has a head-start. The time taken for soil formation can also be complicated by changes in climate, succession of vegetation during soil formation, hydrological changes within the soil during formation, and the possibility of accession of material at the soil surface. There is a strong relationship between the time taken for soil formation and the temperature. Chemical weathering is more rapid in moist tropical environments and this leads to faster and deeper soil formation. Colder climates may have more-rapid physical weathering, for example shattering by frost and erosion by ice, but soils formed on these deposits in cold climates are more weakly developed and shallower. The rate at which soils form and the type of soil can be greatly affected by the nature of the bedrock or deposited sediment. Soils generally form more slowly from silica-rich or quartzo-feldspathic rocks such as greywacke or granite as compared to rocks rich in iron-bearing minerals such as basalt and gabbro. The same is true for sediments and volcanic ash. Silica-rich alluvium weathers more slowly than sediment derived from basalt or from basaltic volcanic ash. The tectonic and physical weathering history of a rock is also important. Any rocks shattered by faulting, folding and uplift, or by physical weathering such as frost action, will weather more quickly to a soil than will solid counterparts. For sediments, texture plays a role. Fine-grained deposits have a much greater surface area for the effects of weathering to act upon.

Scientists intrigued about how rapidly soils form and the transformations that take place have often compared the sequences of soils on the same kind of parent material, the same type of landscape and the same climate. It is usually difficult to escape the influence of changing vegetation with time, for example colonisation of a freshly deposited sediment by lichens, followed by grasses and herbs, then shrubs, trees that are frost-tolerant when young and then frost-sensitive species that use the shelter from frost provided by the preceding trees. Therefore, if one does find a series of similar deposits but of ascending age, the soil-forming factors of climate, relief (landscape) and parent material might be the same but the vegetation seldom is. Such a series is known as a *chronosequence of soils* and may be written:

$$S = f(o, t)cl,r,p$$

where the main soil-forming factor is time (t) of soil formation and its attendant vegetation (o), and the other soil-forming factors of climate (cl), relief or topography (r) and parent material (p) are the same in terms of their influence on soil formation.

We have used the sand dunes that are found along the coast of the south-western part of the North Island as an example when discussing relief or topography (Figure 1.9). The dunes can also be used as an excellent example of a chronosequence. Soil evidence and dune morphology tell us that the sand has accumulated unevenly over time. Between 30,000 and about 15,000 years ago, during the Last Glacial Period when temperatures were as much as 5–7°C lower, it was too cool for forest. The rivers formed vast, gravelly plains because, as described earlier, erosion rates in the mountains and high country were much greater. In places where the rivers deposited finer sandy or silty sediments, it could dry and blow as dune sands, or dust known as loess. Some dunes of this age may also have been blown from the coastline, but at the time the sea was 120 m lower and the coastline 20–50 km west of where it is now. The dunes apparently stabilised when the climate warmed and forest returned to the area about 15,000 years ago, and soil has been forming in earnest since that time. These dunes are collectively called the Koputaroa dunes.

When the continental ice sheets and glacial ice in places like the Southern Alps melted at the end of the Last Glacial, the meltwater poured into the sea and the sea-level began to rise. It did not reach its current level until about 7000 years ago, although, at that time, the coastline in the coastal sand country was several kilometres inland from where it is now. It is thought that as the sea-level rose, dune sands accumulating on the coast were pushed inland and eventually stabilised. The age of these dunes is not known precisely but most are probably 7000–4000 years old. They are

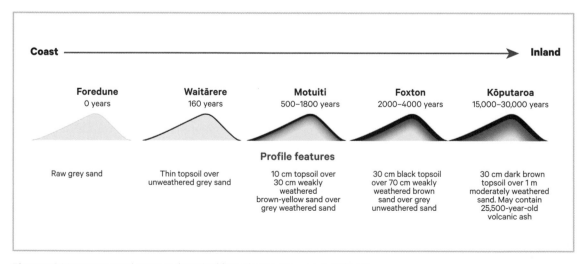

Figure 1.9 Manawatū sand country dune-building phases. (Source: A. Palmer)

known as the Foxton dunes. In most areas, prior to Māori and European occupation these dunes had an open forest cover including coprosmas, kānuka and totara.

Soil and geomorphological evidence suggests that there was then somewhat of a hiatus in dune formation. About 1800 years ago there was a major volcanic eruption from Taupō volcano in the central North Island. Pumice was scattered into the headwaters of rivers draining the central North Island in all directions. Vast quantities were transported down the Whanganui River and, to a lesser extent, the Rangitīkei. Mixing with other quartzo-feldspathic sediment at the coastline, the sediment was pushed south by wave action and longshore drift. The sand supply promoted a new phase of dune-building called the Motuiti dunes. Tiny pieces of Taupō pumice, blown as part of the dune-sand, can be found in these dunes, and lines of sea-rafted pumice often lie on their seaward side. A number of Motuiti dunes appear to have been re-activated by burning during Māori occupation in the district commencing approximately 700 years ago. The dunes were eventually stabilised, and by the time of European settlement had an open cover of shrubs and grasses. Bracken fern was prolific on dunes occupied by Māori.

About 160 years ago European settlers arrived, driving cattle and sheep along the coast to newly acquired land they had purchased for farming. Fire during land clearance and physical disturbance by the cattle resulted in a new phase of sand-dune formation known as the Waitarere dunes. These dunes have gradually been stabilised by introduced and native grasses, shrubs and weeds. Foredunes at the coast, and a few other places where dunes have re-activated, are only partially covered by native grasses, and so sand is still actively blowing in these areas.

A chronosequence of soils has formed on these dunes of increasing age inland.

Describing the soil profile

To understand the origin and formation of the soil and its properties we need to be able to describe the soil with reference to key features of its profile. First, the soil profile is marked off in a series of *horizons*. These are zones of soil sub-parallel to the ground surface that have contrasting properties. There are four major physical properties that are used to describe each soil horizon when the pedologist is in the field. These are soil colour, texture, consistence, and structure. Where any one of these properties changes, a new horizon is drawn.

Soil colour

Soil colour is an important aspect of soils. The common colouring agent in the topsoil is organic matter. These colours may vary considerably depending on the plant cover from which they develop. For example, in the Egmont soils of Taranaki the topsoils formed under forest are generally brown, contrasting with the black topsoils formed under neighbouring coastal scrub. The colour of the soil does not necessarily indicate the amount of organic matter, for a relatively low quantity is present in some black soils.

The common colouring agents in the subsoil are iron oxides. In well-drained soils, these range in colour from red through brown to yellow,

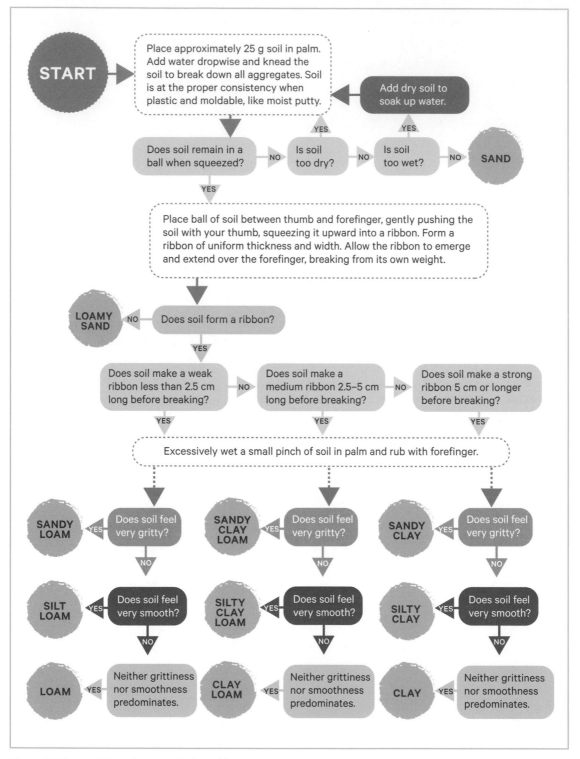

Figure 1.10 Determining soil texture. (Adapted from Thien 1979)

depending in part on the degree of hydration. In many places, subsoil colours reflect not only the nature of the iron oxides but also the amounts inherited from the parent materials. For example, soils from claystone (which has a low iron content) are commonly pale, whereas those from basalt (rich in iron) are commonly red or strong brown. The colours of some soils, especially youthful ones, may not be due to soil-forming processes at all, but may merely reflect the colours of the parent material.

A white or grey layer beneath the surface organic horizon, such as occurs in Podzol soils (soils formed under forested landscapes on coarse parent material that is high in quartz) generally indicates intense leaching. It is commonly associated with strong colours in the subsoil, which indicates illuviation of iron and humus. The colour of quartz grains also yields much information. In eluvial horizons from which iron, clay, and other materials are being removed, and particularly in the presence of acid humus, they are commonly bleached clean, and glisten. In contrast, in illuvial horizons they are coated and dull.

Reddish brown and yellow mottles are nearly always present in imperfectly and poorly drained soils, especially in horizons where the water-table fluctuates. In the presence of organic matter the proportion of grey generally increases with increasing wetness. Where the soils are very grey they are said to be strongly gleyed.

Soil texture

Soil texture is a measure of the proportions of sand, silt and clay within a soil. It refers to that portion of the soil which is less than 2 mm in diameter. Soil texture controls how soils can be utilised, so is therefore an exceptionally important physical property. Soil texture is one of the more difficult physical properties to learn how to assess (Figure 1.10) and requires much practice. Note that in soil science:

- sand = 0.06–2 mm
- silt = 0.002–0.06 mm
- clay < 0.002 mm.

The impediment to land use caused by particles greater than 2 mm in diameter depends on their size. Whereas a moderately gravelly soil might be cultivated, even a slightly stony soil may be difficult to cultivate without causing damage to equipment. See Figure 1.10 for a procedure for determining by feel.

Consistence

Consistence is an assessment of cohesion, adhesion and resistance to deformation of soil under an applied stress. Consistence is influenced by texture, water content, clay mineral composition and organic matter content. The strength of an undisturbed soil sample at the water content encountered on the day of description is determined as the resistance to crushing of a 30 mm cube, or smaller aggregate, of soil. Where natural aggregates are greater than 30 mm, only aggregate strength is recorded; where aggregates are present and are less than 30 mm across, both soil and aggregate strength is recorded; and where no structure is present, only soil strength is recorded. Plasticity (cohesion) and stickiness (adhesion) are measured on moistened, thoroughly remoulded soil, free of roots, concretions and coarse fragments.

Soil structure

Soil structure is the naturally occurring arrangement of soil particles into aggregates that result from pedogenic (soil-forming) processes. One definition of soil structure is the arrangement

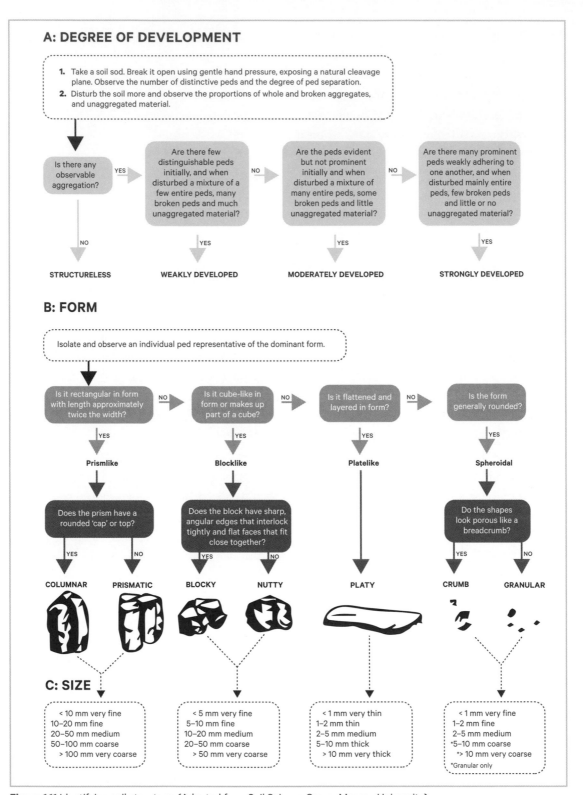

Figure 1.11 Identifying soil structure. (Adapted from Soil Science Group, Massey University)

and organisation of soil particles into aggregates (or peds) *and* the stability of such arrangements to stress. Soil structure is defined in terms of degree of development, form, and size.

The flow diagram for the assessment of soil structure (Figure 1.11) illustrates many of these aspects.

Thien, S.J. 1979. 'A flow diagram for teaching texture-by-feel analysis.' *Journal of Agronomic Education* 8 (1): 54–55.

Trustrum, N.A., and de Rose, R.C. 1988. 'Soil depth–age relationships of landslides on deforested hillslopes, Taranaki, New Zealand.' *Geomorphology* 1: 143–60.

Summary

New Zealand relies on its soils for food production and economic wellbeing. Soil is one of our most valuable natural resources. For such a small country, New Zealand has a wide variety of soils, some of which pose a challenge to use and require careful management if they are to be farmed in a sustainable manner. By understanding how they form and are distributed across landscapes and by appreciating their basic features and properties, we can improve soil management. In other words, knowledge of how soils form and their properties are of paramount importance for the development of good soil management practices.

References

Isaac, M.J. 2015. 'Overview, Maps that Changed New Zealand.' In: Graham I. J., ed. *A Continent on the Move*. 2nd ed. Wellington: Geoscience Society of New Zealand, 25–29.

McLaren, R.G., and Cameron, K.C. 1996. *Soil Science: Sustainable Production and Environmental Protection*. Oxford: Oxford University Press.

Molloy, L., and Christie, Q. 1988. *Soils in the New Zealand Landscape: The Living Mantle*. Wellington: Mallinson Rendel Publishers.

Chapter 2

Pasture and Forages

Lydia Cranston

Chapter 2

Pasture and Forages

Lydia Cranston

School of Agriculture and Environment, Massey University

Introduction

In New Zealand, livestock farming systems are pasture-based, with the main feed source being grazed pasture. Sheep, cattle and deer graze pasture outdoors all year round. The combination of our soil types, mild temperatures and relatively consistent rainfall means that pasture can grow year-round and animals can withstand the external environment. New Zealand has approximately 13.9 million hectares (ha) of pastoral land of which roughly 63 per cent is used for sheep and beef cattle farming, 18 per cent for dairy farming and the remainder for cropping and other farming enterprises (Statistics New Zealand 2017).

New Zealand livestock farming systems are typically lower cost than most overseas systems in which livestock are housed indoors for all or part of the year and a large proportion of the diet comes from processed foods or grain. Moreover, in contrast to most other countries,

New Zealand has a very small population relative to its agricultural production and therefore the domestic market consumes only 5 per cent of total milk production, 10–20 per cent of total beef production and 3–5 per cent of total sheep meat production.

On a global scale, New Zealand accounts for approximately 30 per cent of all dairy products, 6–8 per cent of all beef and 50 per cent of all lamb meat traded internationally. Agriculture is a major contributor to the New Zealand economy, representing 62 per cent of New Zealand's export earnings in 2017 (Statistics New Zealand 2018).

The practice of farming in New Zealand is not subsidised by the government, unlike many overseas systems where subsidies keep farming economically viable. As a result, New Zealand farmers are encouraged to run efficient systems by maximising the use of home-grown pasture in order to maintain profitability. Further, the number of sheep, beef cattle and dairy cattle and the area of land used for growing horticultural crops varies from year to year depending on the relative prices attainable in each of the sectors.

The history of bush to pasture

When European settlers arrived in New Zealand more than 200 years ago most of the country was covered in forest (largely native podocarp and broadleaf trees). Large trees were felled for timber, and fires were used to clear bush and create ground suitable for growing food crops and pasture. The original pasture seed mixes included browntop (*Agrostis capillaris*), perennial ryegrass (*Lolium perenne*), cocksfoot (*Dactylis glomerata*), crested dog's-tail (*Cynosurus cristatus*), white clover (*Trifolium repens*), red clover (*Trifolium pratense*), subterranean clover (*Trifolium subterraneum*) and lotus (*Lotus pedunculatus*). The seed was broadcast by hand at rates of 20–40 kg/ha. Over time, the introduction of artificial fertilisers, paddock subdivision and soil cultivation has enabled more-productive pasture mixes, predominantly perennial ryegrass/white clover, to be sown on large areas of pastoral land.

Pastoral land that was once covered in bush.

Pasture production

Pasture production and species composition are determined by four key factors: temperature, soil moisture content, soil fertility and grazing management (Scott et al. 1985). Annual pasture production in New Zealand ranges between approximately 6 and 18 tonnes of dry matter (tDM) per hectare per year. Dairy farms are generally located on flat land in high-rainfall environments with high soil fertility, and typically grow at least 12 tDM/ha/year. Lowland finishing farms are generally located on flat to rolling land, but generally have slightly lower soil fertility levels and fertiliser inputs and so typically grow between 9 and 12 tDM/ha/year. Hill-country sheep and beef farms are often located in dry-summer or cold-winter environments, have low fertiliser inputs, and therefore typically grow between 6 and 10 tDM/ha/year.

The effect of climate on pasture production and species

Plant species have different temperature requirements; in temperate pasture species, temperatures below 5–10°C or above 25°C generally restrict plant growth. Similarly, soil moisture content (either restriction or waterlogging) can limit plant growth. The combination of temperature and moisture helps to determine both the annual production and the seasonal distribution of production.

In New Zealand there are four general patterns of pasture production. These represent the varying climates across different regions (Scott et al. 1985; Figure 2.1). In Northland, with its warm, wet environment, annual pasture production is greater than in other climate regions. High-yielding subtropical pasture species such as

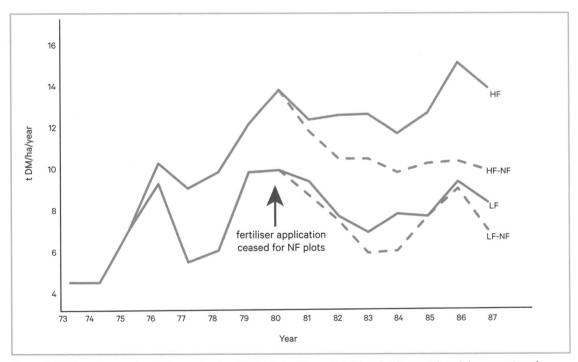

Figure 2.1 Variation over a calendar year in annual pasture production and seasonal pasture production with temperature and moisture. (Adapted from Scott et al. 1985)

Figure 2.2 Effect of superphosphate application (HF, high fertiliser rate; LF, low fertiliser rate) and the cessation of superphosphate application (NF, no fertiliser) on the pasture production of North Island hill-country farm Ballantrae. The variation in pasture production varies from year to year depending on annual rainfall and temperature patterns, however, the long-term trends indicate a clear positive response between the rate of superphosphate application and annual pasture production. (Adapted from Lambert, Clark and Mackay 1990)

kikuyu (*Pennisetum cladestinum*) and paspallum (*Pasp-alum dilatatum*), which have greater growth rates at higher temperatures, thrive in and dominate this region.

Temperate pasture species are prevalent in other parts of New Zealand. In Southland, a cold, wet environment, pasture growth is restricted during winter; in summer, regular rainfall and moderate temperatures enable optimum growth and quality of forage. This region is particularly suitable for growing young stock and summer milk production.

In Wairarapa, a warm, dry environment, pasture growth generally ceases during winter due to low temperatures. During spring, pasture growth rates soar with optimal temperatures and soil moisture; in summer, pasture growth is restricted due to high temperatures and low soil-moisture levels. In this environment, farmers must have flexible livestock policies whereby livestock numbers can be increased or decreased quickly based on pasture availability.

In Central Otago, a cold, dry environment, pasture growth is severely restricted for large parts of the year due to low winter temperatures and high summer temperatures. This results in seasonal pasture growth with periods of growth in spring and autumn. This region has the lowest annual pasture production compared with other climatic regions, making farming in this environment very challenging.

The effect of soil fertility and pH on pasture production and species

Different plant species have different fertility requirements. Lower-fertility species like browntop, crested dog's-tail and lotus dominate hill country, while higher-fertility species like pe-rennial ryegrass and white clover are more commonly found on flat land with more fertile soils. The pasture species that dominate in lower-fertility environments typically have lower annual pasture production and poorer nutritive values compared with pasture species found in higher-fertility environments. As a result, hill-country pasture usually supports less livestock per hectare compared with flat land, and therefore produces lower economic returns per hectare.

Phosphorus and nitrogen are the primary soil nutrients that can limit plant growth. Artificial fertiliser containing these nutrients can be applied to soils to help increase their natural fertility level and increase pasture production. New Zealand began importing phosphate fertiliser in 1867 in the form of rock phosphate. The bulk of the rock phosphate imported into New Zealand now comes from Morocco, Western Sahara and South Africa. In New Zealand, much of the rock phosphate is processed further by adding sulphuric acid to create superphosphate. Superphosphate breaks down quickly in the soil, enabling a faster plant growth response compared with rock phosphate, and is now the most commonly used fertiliser in New Zealand. A long-term fertiliser trial on a North Island hill-country farm, Ballantrae, demonstrated the benefits of applying superphosphate, with annual pasture production increasing by upwards of 5 tDM/ha depending on the application rate, and declining when annual applications of superphosphate ceased (Lambert Clark and Mackay 1990; Figure 2.2).

Nitrogen fertiliser is manufactured by converting atmospheric nitrogen to ammonia and then urea. Nitrogen fertiliser can increase pasture production by 1–2 tDM/ha and is most commonly applied in spring and autumn when pasture growth is commonly limited by soil nitrogen levels. The extra DM grown as a result of the nitrogen fertiliser application is a cheaper form of additional feed when compared with feeds that can be bought in, such as hay, silage or grain. Annually, dairy farms commonly apply between 50 and 150 kg N/ha across multiple applications in spring and autumn to boost cow milk production. Sheep and beef farms typically apply less nitrogen fertiliser because their economic returns per hectare are lower; they use targeted applications of nitrogen fertiliser when total farm pasture supply is below typical levels following a poor growing season, rather than set annual applications.

Soil pH has a significant effect on plant growth. Pasture growth is optimised when soil pH is between 5.6 and 6.2. Many pasture species, particularly legumes, will not grow or will yield poorly when the soil pH is outside this range. Furthermore, soil pH has an interactive effect on the availability of phosphorus in the soil. In acidic soils (with a pH <5.6), an increasing amount of phosphate is bound to soil particles and unavailable for plant use, which limits growth.

Over time, and with the application of nitrogen and phosphate-based fertilisers, soil pH naturally becomes more acidic. Lime can be applied to soils to increase soil pH; 1 tonne of lime per hectare increases soil pH by approximately 0.1 units. Nitrogen and phosphate fertilisers are commonly applied annually, while lime is generally only applied every 5–15 years when soil pH lowers sufficiently to reduce pasture production.

On flat land fertiliser and lime are applied by truck, but on undulating and steep land fertiliser is more commonly applied by aeroplane. The cost of application is significantly higher with aerial application. Consequently, hill country is typically fertilised less frequently and at lower application rates compared with flat land, resulting in soils with lower fertility status in hill country. Further, lime is considerably denser than phosphate-based fertiliser and therefore much more expensive to apply. The cost of applying lime on hill country is high, and consequently many hill-country farmers have acidic soils with poor rates of pasture production.

The effect of grazing management on pasture production and species

Grazing management can limit pasture production via two mechanisms.

- Under-grazing allows dead material to accumulate and shade pasture, thereby reducing the rate of photosynthesis.
- Over-grazing reduces pasture density and exposes bare areas, thereby reducing the total leaf area available for photosynthesis.

The common saying 'grass grows grass' refers to the effect that greater above-ground plant material enables greater rates of photosynthesis and therefore pasture growth.

The timing and frequency of grazing also affects pasture production. In general, plants should be grazed when they have sufficient carbohydrate reserves in their roots to maximise future growth and plant persistence. The optimum grazing regimen differs between pasture species, but typically ranges between 3 and 6 weeks between grazing events.

Ryegrass

Ryegrass is the most widely used pasture genus in New Zealand because it is fast to establish, has a high yield and nutritive value, is tolerant

A grass pasture ready to graze on left and recently grazed on right.

of hard grazing and is compatible with clover. There are many species of ryegrass, which vary in DM yield and winter growth, nutritive value and persistence. Perennial ryegrass is the most commonly sown ryegrass species and will persist for between 5 and 20 years. Italian ryegrass (*Lolium multiflorum*) has larger leaves, is higher-yielding and has a greater proportion of its growth in winter (winter-active), but only persists for 1–3 years. Annual ryegrass is extremely winter-active and high-yielding relative to perennial ryegrass, but it only persists for 6–12 months. Ryegrass is naturally a diploid plant, meaning that it has two sets of chromosomes. Plant breeders have modified some ryegrass varieties to be tetraploid (have four sets of chromosomes). Tetraploid varieties are faster to establish, have larger leaves and greater DM production, and have a higher sugar content which makes them preferable to livestock; however, they are less persistent.

Ryegrass contains a naturally occurring fungus (an endophyte; *Neotyphodium lolii*) which lives in the plant and produces alkaloids (peramine, ergovaline, lolitrem B). These alkaloids help deter insect pests but also cause ryegrass staggers (a nervous disorder), which reduces animal production. Plant breeders have created novel endophytes that produce alternative alkaloids (e.g. Epoxy-janthitrems) which have minimal effect on animal health while still protecting the plant from insect attack (Thom et al. 2006).

When farmers choose to establish a new ryegrass pasture, they must choose a variety that is suitable for their climate, livestock class and farm system based on DM yield, degree of winter growth, persistence, ploidy level and endophyte type.

The value of legumes

Grass species are commonly sown in a mixed sward with a legume species, usually clover. Most mixed pastures contain between 5 and

20 per cent legume content, with the legume content rarely exceeding 30 per cent (Stewart and Charlton 2006).

Legumes are of great importance in temperate grasslands and have two main functions. First, they fix nitrogen from the atmosphere. They utilise some of this nitrogen to support their own growth, but also generate excess nitrogen that can be used by other plant species to support their growth. The total amount of nitrogen fixed in a given pasture sward can range from 0 to 350 kg N/ha/year, and is dependent on the proportion of legume in the pasture, legume growth rate, soil nutrient content, and soil temperature and moisture (Hoglund et al. 1979). Second, legumes have a higher protein and metabolisable energy content compared with grass species and therefore enhance the overall nutritive value of the pasture (Figure 2.3). This helps to drive

animal production, with swards with greater legume content supporting higher rates of live weight gain (McEwan et al. 1988). Furthermore, the pattern of seasonal pasture production for legumes is complementary to that of grass, with legumes growing more during late spring and summer when grass growth is typically restricted due to high temperatures and moisture stress, and less during winter when grass is in a leafy state and therefore has a high nutritive value (Figure 2.3).

The most predominant legume found in New Zealand pastures is white clover (*Trifolium repens*). White clover is tolerant of hard grazing, grows in a wide range of temperature and fertility environments, and fixes a large amount of nitrogen (Stewart and Charlton 2006). However, it has shallow roots and a low root to shoot ratio, and is therefore not very tolerant of

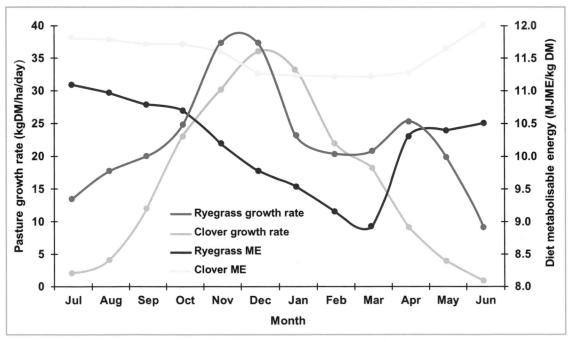

Figure 2.3 Annual distribution of pasture production and metabolisable energy (ME) content of perennial ryegrass and white clover growing separately. (Adapted from Harris and Hoglund 1977)

dry conditions (Brock and Hay 2001). In drier environments, subterranean clover (*Trifolium subterraneum*) and red clover (*Trifolium pratense*) are more suitable and can make a more significant contribution to the pasture sward. Subterranean clover is an annual species which germinates and grows vigorously during spring under optimal growth conditions, and then sets seed in the soil in summer before dying off. Red clover is a perennial species with a taproot which enables it to extract moisture from deep within the soil profile, and to continue to grow during dry summer months. In lower-fertility hill-country environments lotus can be the dominant legume species in the pasture.

Pasture establishment

There are two main methods of establishing pastures: cultivation and no-tillage (direct-drilling).

- Traditionally, all new pastures were established using cultivation, which involves spraying out old pasture, turning the soil over, and breaking up the soil to form a fine seedbed. This method results in a consistent seedbed and generally a successful pasture establishment, however, it is very time-consuming, and regular cultivation can lead to damage to the soil profile.
- Alternatively, no-tillage involves spraying out old pasture and then slicing the ground and directly sowing rows of seeds into the existing soil structure. This method is much faster than cultivation; however, the success of the establishment can be more variable. The seedlings may have to establish in less than optimal conditions, including variable

soil moisture, shading from the preceding plant species (dead material or growth from plants which failed to die following spraying) and greater competition from weed species.

New pastures can be established in either spring or autumn, when temperatures and soil moisture levels allow for successful seed establishment and growth. Re-sowing during autumn generally results in grass pastures with lower contributions of weeds compared with spring-sown pastures. In spring, weeds have a similar rate of germination and establishment to grass species such as perennial ryegrass; in autumn, temperature conditions are less favourable for weed germination whereas perennial ryegrass has been bred for fast establishment in cooler conditions. Many farmers spray out existing pasture in spring, sow a short-term summer-active brassica (e.g. turnips; *Brassica rapa*) or herb crop (e.g. chicory; *Cichorium intybus* or plantain; *Plantago lanceolata*) for grazing over summer, and then spray and establish a grass pasture in autumn. This 'double spraying' helps to remove hard-to-kill weeds and allows for a more successful establishment of a perennial grass pasture.

Pasture sowing rates vary between different species depending on the size of the seeds, method of establishment and the site of establishment (Table 2.1). Pasture species with large seeds require higher sowing rates compared to small-seed species. Pastures established by cultivation require higher sowing rates relative to no-tillage cultivation, as seeds are sown in rows rather than scattered across the entire area. Sites with less than optimal germination conditions may require higher sowing rates to ensure that the sown species has sufficient plant numbers to enable high yields and

A newly established perennial ryegrass and white clover sward.

Table 2.1 Seed weights and sowing rate ranges of the major grass and legume species for use in compiling seed mixtures.

Species	Seeds/kg	Sowing rate range in mixtures (kg/ha)
Grasses		
Perennial ryegrass	500,000	10–25
Cocksfoot	1,000,000	4–10
Tall fescue	400,000	16–20
Yorkshire fog	3,300,000	1–5
Browntop	12,500,000	1–2
Legumes		
White clover	1,400,000	3–5
Red clover	295,000	4–6
Subterranean clover	150,000	2–6
Lotus	1,200,000	1–5

out-compete weed species. Sowing pasture mixtures with a high grass sowing rate will result in pastures with lower legume contribu-

tions, as legumes are slower to establish than grass species. However, these swards will have a denser ground coverage during the establishment phase and will be less vulnerable to weed invasion, which is advantageous in sites with significant weed issues, and may be more persistent in the long term. However, swards with lower legume plant numbers will have lower levels of nitrogen fixation and reduced animal performance compared to swards with a higher legume plant population.

Matching pasture production and animal feed demand

The aim of a pastoral farming system is to match the pattern of animal feed demand with the pattern of pasture production. The pattern of pasture production differs between regions based on their climate (Figure 2.1); however, in general, pasture production peaks during spring, is restricted during summer due to warm temperatures and soil moisture deficit, increases during autumn as soil moisture levels rise and is restricted during winter due to low temperatures (Figure 2.3).

Livestock give birth during spring when pasture growth is at its fastest. Livestock numbers are then reduced throughout summer and autumn, by selling young stock and culling unproductive and old livestock, to ensure that a minimum number of livestock remain on farm during winter. However, there is large year-to-year variation in seasonal pasture production and therefore forage crops and other supplementary feeds are often utilised to ensure that animal feed demand is met.

The following is a seasonal calendar of key pasture growth and management.

Spring

Animal feed demand reaches a peak as breeding animals give birth and require significantly higher feed intakes to support lactation. Calving and lambing dates are timed to coincide with peak spring pasture production, and the growth rates of finishing animals are increased to consume more of the available pasture.

Pasture production during spring commonly exceeds animal demand. Farmers can utilise this extra feed by allowing the pasture cover to build up in some paddocks and saving this feed to use during summer (standing hay), or harvesting the pasture to make hay or silage. Alternatively, farmers can reduce the area of pasture available for grazing by spraying out paddocks and sowing a forage crop for use in summer or winter or re-sowing in a grass/clover mix. This helps by shifting the pattern of feed supply to a time when pasture production or quality is low, to allow it to better match animal feed demand. The remaining available grazing area will then have an increased stocking rate, which will help to reduce stem production and maintain pasture quality going into summer. Annual applications of phosphate fertiliser are commonly applied during spring to time with peak pasture growth. Nitrogen fertiliser can also be applied to increase pasture production to help match peak animal demand or allow hay or silage to be made.

Summer

During summer, pasture quality is typically poor as grass species undergo reproductive growth, producing more stem and seed-head material as opposed to leafy growth. Furthermore, pasture production is often limited due to high temperatures and soil moisture deficits. Some farms utilise summer-active forage crops such as lucerne

Lambs grazing a mix of chicory, plantain and clover during summer.

Ewes and rams grazing dry pasture during late summer/early autumn.

(*Medicago sativa*), chicory or turnips to increase the amount of home-grown summer feed. These forage crops supply high-quality feed, in terms of both energy and protein, and are commonly used for growing young stock (lambs or weaner calves) or for supporting greater rates of milk production in dairy cows.

In some years, drought can severely affect pasture production and require farmers to sell breeding-age livestock in addition to young stock. This has a longer-term effect on the farm system, as reduced livestock will be born in subsequent seasons until replacement animals reach breeding age; breeding livestock may have to be bought in from other farms at considerable cost.

Autumn

During autumn pasture growth rates increase following autumn rains. Farmers commonly allow pasture cover levels to build up prior to the start of winter, as a way of transferring feed into winter (when low pasture growth rates occur). Further, livestock numbers are adjusted during autumn by selling livestock, drying off cows or sending livestock away for grazing on other properties. This means that livestock numbers are reduced to a minimum by the start of winter. Nitrogen fertiliser can also be applied in autumn to boost pasture growth rates and increase the amount of feed available for winter.

Winter

During winter animal-feed demand is low; however, pasture growth rates are generally insufficient to meet the total animal-feed demand across the farm. Animal intakes are generally restricted by strip-grazing pasture, with supplementary feeding in the form of hay, silage or

Young cattle strip-grazing forage oats behind a temporary electric fence.

Sheep strip-grazing swedes during winter.

winter forage crops being used to ensure that animals are offered maintenance feed requirements. The metabolisable energy content of hay and silage is generally 10–20 per cent lower than fresh grass pasture and it is therefore difficult for an animal to gain weight over winter when these supplements contribute substantially to the diet. Similarly, winter forage crops generally have a high water content, which can limit animal intake and thus potential live weight gain.

Winter forage crops include brassica species, such as swedes (*Brassica napobrassica*) and kale (*Brassica oleracea*), forage oats (*Avena sativa*) or fodderbeet (*Beta vulgaris*). These are sown in the previous spring, summer or autumn and allowed to grow and build up large DM yields during this time. They are capable of growing between 5 and 30 tDM/ha depending on the environment and forage type, and commonly achieve yields of 10–15 tDM/ha. Consequently, these forage crops are commonly grazed using a high stocking rate of animals behind a temporary electric fence, with a fresh allocation of crop offered to the animals at regular intervals. Brassica species and fodderbeet are low in fibre, and are therefore often fed in conjunction with hay or silage to meet animal requirements.

Alternative pasture species

Perennial ryegrass/white clover is the most commonly sown pasture mixture in New Zealand. However, as previously discussed, its growth rate is often restricted during warm, dry summer conditions and cold, wet winters. Alternative pasture species that are either summer-active (chicory, plantain, red clover) or grow during spring, summer and autumn and are fed during winter (winter crops; oats, swedes, kale, fodderbeet) are used to help fill feed gaps. There has been considerable development of new cultivars of these alternative species since the 1990s, including forage chicory, plantain, red clover, raphnobrassica and fodderbeet. The advantages that these alternative forages have over grass species include improved feed quality and higher DM yields during either summer or winter. Consequently, compared with grass they can support greater animal performance, including higher weight gains and greater milk production, especially during summer.

Chicory and plantain cultivars are very palatable to livestock and have been bred for upright growth structure and longer persistence. Further, plantain has been shown to have positive environmental effects, including reducing nitrogen leaching (Navarrete et al. 2018). The use of these forages is continuing to increase across New Zealand as both specialist crops and in grass mixtures. Chicory and plantain generally yield a similar amount of dry matter to perennial ryegrass (10–14 tDM/ha), but a greater proportion of the growth occurs during spring and summer and less during winter. Further, as climate change continues to increase average temperatures, perennial ryegrass will become less suitable in some places within New Zealand,

A herb and clover mix containing chicory, plantain, red clover and white clover.

Lambs grazing raphnobrassica (a kale × radish hybrid) during summer.

and alternative species may be increasingly used. Herb and clover mixes containing chicory, plantain, red clover and white clover are increasingly being sown by farmers because they can last for 3–4 years under rotational grazing, which is longer than single-species crops.

Red clover was traditionally considered a biennial plant and was grown as a monoculture for grazing lambs over summer or for making hay. However, modern cultivars have been bred to be more persistent under rotational grazing in grass mixtures (Ford and Barrett 2011). Today, red clover is more commonly included in perennial ryegrass and white clover mixtures, thereby increasing summer feed quality and nitrogen fixation. Further, red clover is more tolerant of clover root weevil compared with white clover (Ferguson, Barton and Philip 2016), and is therefore particularly important for including in pasture mixes in environments where clover root weevil is prevalent.

Raphnobrassica is a kale × radish hybrid that was released commercially in 2018 as a multiple-graze plant bred for greater drought tolerance, and improved disease and insect tolerance compared with older kale cultivars. Raphnobrassica can be grazed by livestock over summer with the paddock then shut up to allow regrowth for winter grazing (dual-season crop).

Other winter-forage crops such as swedes, fodderbeet and kale have been available for many years and are commonly used throughout New Zealand. Single-graze crops such as swedes and fodderbeet produce higher DM yields available for winter feed (10–30 tDM/ha) than multiple-graze crops (kale, raphnobrassica), which yield 5–6 tDM/ha for summer feed and 5–10 tDM/ha for winter feed. Swedes are typically used for winter sheep feed in the colder parts of New Zealand (Southland and Central North Island), while kale and fodderbeet are more commonly fed to finishing cattle and dairy cows. Fodderbeet tends to be more energy-dense than brassica species and the use of it has increased significantly throughout the 2000s.

Environmental effects of pastoral farming

Since the 1980s, when government subsidies were removed, pastoral farming in New Zealand has continued to intensify, with farmers increasing the numbers of livestock per hectare in order to increase their economic returns. This

was further extenuated by the expansion of the dairy industry in the 1990s. This intensification means that there are greater potential nutrient losses of nitrogen (N) and phosphorus (P) from drainage and run-off (Field 1985, Field et al. 1985; Ball and Theobald 1985; Ruz-Jerez, White and Ball 1995). High levels of N and P are considered to be pollutants in waterways, resulting in negative environmental effects and human health hazards (Di and Cameron 2002).

Both nationally and regionally, government is beginning to place restrictions on maximum N and P losses from livestock farms in order to improve surface and groundwater quality. These restrictions are determined on the basis of soil type (as some soil types are more likely to leach nutrients than others), climate (amount of rainfall/irrigation applied) and proximity to waterways.

Wintering livestock on crops is coming under scrutiny, as high stocking rates can cause significant mud development, damage soil structure, and result in a high concentration of nutrients across a small area (from dung and urine), thereby increasing the likelihood of nutrient leaching.

Farmers are being encouraged to carefully select their paddocks for winter cropping based on soil type, slope and proximity to waterways. Further, in higher-risk paddocks it is advised to graze with lighter livestock such as sheep or young cattle, and to utilise grass buffer zones around waterways and at the bottom of slopes. In addition to improved winter cropping practices, many farmers are fencing off waterways from livestock and utilising innovative farm practices, including changing fertiliser application rates and the methods and timing of fertiliser application, sowing alternative pasture species, and lowering stocking rates in order to reduce N and P loss.

Alternative land use

Mānuka (*Leptospermum scoparium*) is a native New Zealand plant that readily colonises bare or steep land and was historically considered a weed. However, in recent years the price of mānuka flower honey has soared because of its antibacterial properties. In 2017, the price of mānuka honey ranged from $12 to $135 per kg, up from $10.45–$60 per kg in 2013 (Ministry for Primary Industries 2018). As a result, some steeper hill country has been allowed to revert from pasture to mānuka, while in other places mānuka has been planted on farms to enable farmers to take advantage of an alternative revenue stream.

Pasture pests

Insects
Grass grub (*Costelytra zealandica*), porina (*Wiseana cervinata*, *W. copularis*), Argentine stem weevil (*Listronotus bonariensis*) and clover root weevil (*Sitona lepidus*) are the most significant insect pasture pests in New Zealand. All of these insects feed on pasture (above or below ground) at one time during their life-cycle. They can be particularly damaging during pasture establishment but can also cause significant damage to older pastures, resulting in reduced persistence.

Grass grub beetles lay eggs in the soil, and larvae hatch during spring and feed on the roots of pasture species. This root damage causes affected plants to have reduced above-ground growth and to be more vulnerable to being removed during grazing, resulting in reduced pasture persistence. They are most damaging during autumn, when pasture production can be reduced by up to 40 per cent. The damage can be observed as patches of yellow grass within a paddock, and is particularly visible in late winter to early spring. The larvae show preference for feeding on particular pasture species, notably white clover, perennial ryegrass and plantain.

Porina moths fly during spring and early summer and lay eggs in pasture. Caterpillars hatch from eggs and live in vertical tunnels in the soil, up to 30 cm deep. The caterpillars emerge from their tunnels at night and graze pasture. Their damage is most noticeable during late autumn to early winter and is observed as bare patches in the soil. A pesticide can be sprayed on to paddocks during autumn to help control porina populations and reduce pasture damage.

Argentine stem weevils (ASW) are only 3–4 mm long. The larvae graze the growing points of tillers, resulting in tiller death. Large infestations can dramatically reduce pasture persistence and can be devastating in establishing pastures. In 1990, a parasitoid wasp was introduced to control ASW; however, over time ASW have evolved a degree of resistance to this wasp. Novel endophytes are now the main method for controlling ASW in perennial ryegrass, with the alkaloids produced deterring the larvae from grazing the plants.

Clover root weevil can cause significant damage to clover populations. The adult-stage weevil feeds on clover leaves (distinctive notching is observed when present), while the larval-stages feed on clover roots. This results not only in reduced clover growth but also in reduced nitrogen fixation and thus reduced grass growth. Red clover is more tolerant of clover root weevil than white clover (Ferguson, Barton and Philip 2016) and it is therefore advisable to include both red and white clover in new pasture mixes. A parasitoid wasp was introduced to New Zealand in 2006 and has been successfully used to help suppress clover root weevil populations.

Other pests

Wild rabbits and possums are distributed throughout New Zealand and are significant agricultural pests. They graze pasture and reduce the total amount of pasture available for farmed livestock. If left uncontrolled, their populations can multiply quickly and can therefore limit livestock stocking rates. Further, possums are carriers of bovine tuberculosis and are therefore a concern for the cattle industry. A national organisation (OSPRI) is responsible for controlling possum numbers and preventing the spread of bovine tuberculosis.

Conclusions

Pastoral farming of livestock has been practised in New Zealand since the early 1800s. Over time there have been significant improvements in our farming systems. These include more appropriate selection and breeding of pasture species, greater understanding of the patterns of pasture production in different regions, and the effect of grazing management and an increased application of fertilisers. There have also been numerous challenges along the way: pests reduce pasture production, and higher stocking rates that increase economic returns also increase nutrient leaching and thereby reduce water quality. In the future we are likely to see climate change drive a shift in our pastoral plant breeding strategies, with more emphasis on species that are tolerant to heat and drought stress, as well as an increased focus on pasture species that might reduce our environmental footprint. Further, as the number of pastoral species and varieties increases, there will be greater need for agronomic advisors who have a broad understanding of pastoral species, soil properties and fertility as well as different livestock production systems.

References

Brock, J., and Hay, M. 2001. 'White clover performance in sown pastures: A biological/ecological perspective.' *Proceedings of the New Zealand Grassland Association* 63: 73–84.

Di, H., and Cameron, K. 2002. 'Nitrate leaching and pasture production from different nitrogen sources on a shallow stony soil under flood-irrigated dairy pasture'. *Australian Journal of Soil Research* 40 (2): 317–34.

Ferguson, C.M., Barton, D.M., and Philip, B.A. 2016. 'Clover root weevil tolerance of clover cultivars.' *Proceedings of the New Zealand Grassland Association* 78: 197–202.

Field, T.R.O., Ball, P.R., and Theobald, P.W. 1985. 'Leaching of nitrate from sheep-grazed pastures.' *Proceedings of the New Zealand Grassland Association* 46: 209–214.

Field, T.R.O., Theobald, P.W., Ball, P.R., and Clothier, B.E. 1985. 'Leaching losses of nitrate from cattle urine applied to a lysimeter.' *Proceedings of the Agronomy Society of New Zealand* 15: 137–41.

Ford, J.L., and Barrett, B.A. 2011. 'Improving red clover persistence under grazing'. *Proceedings of the New Zealand Grassland Association* 73: 119–24.

Harris, W., and Hoglund, J.H. 1977. 'Influences of seasonal growth periodicity and N-fixation on competitive combining abilities of grasses and legumes.' *Proceedings of the XIII International Grassland Congress* 239–243.

Hoglund, J.H., Crush, J.R., Brock, J.L., Ball, R., and Carran, R.A. 1979. 'Nitrogen fixation in pasture. XII. General discussion.' *New Zealand Journal of Experimental Agriculture* 7 (1): 45–51.

Lambert, M.G., Clark, D.A., and Mackay, A.D. 1990. 'Long term effects of withholding phosphate application on North Island hill country: Ballantrae.' *Proceedings of the New Zealand Grassland Association* 51: 25–8.

McEwan, J.C., McDonald, R.C., Fennessy, P.F., Thompson, K.F., Hughson, C.G., and Dodds, K.G. 1988. 'Pasture quality and nutrition in weaned lambs.' *Proceedings of the Society of Sheep and Beef Cattle Veterinarians of the New Zealand Veterinary Association* 18: 178–88.

Ministry for Primary Industries. 2018. Apiculture Monitoring Report. Accessed 25 August 2020. www.mpi.govt.nz/dmsdocument/34329-apiculture-monitoring-report-2018.

Navarrete, S., Kemp, P.D., Rodriguez, M.J., Hedley, M.J., Horne, D.J., and Hanly, J.A. 2018. 'Milk production from plantain (*Plantago lanceolata* L.) pastures.' Proceedings of the Australasian Dairy Science Symposium: 6–11. Accessed 25 August 2020. https://australasiandairyscience.com/wp-content/uploads/2020/03/2018-ADSS-proceedings.pdf.

Ruz-Jerez, B.E., White, R.E., and Ball, P.R. 1995. 'A comparison of nitrate leaching under clover-based pastures and nitrogen-fertilized grass grazed by sheep.' *Journal of Agricultural Science* 125: 361–69.

Scott, D., Keoghan, J.M., Cossens, G.G., Maunsell, L.A., Floate, M.J.S., Wills, B.J., and Douglas, G.B. 1985. 'Limitations to pasture production and choice of species.' In *Using herbage cultivars*, Grassland Research and Practice series No. 3, edited by R.E. Burgess and J. L. Brock, 9–15. Palmerston North: New Zealand Grassland Association.

Statistics New Zealand. 2017. Agricultural production Statistics: June 2017 (Final). Accessed 25 August 2020. www.stats.govt.nz/information-releases/agricultural-production-statistics-june-2017-final.

Statistics New Zealand. 2018. Gross domestic product by industry. Accessed 25 August 2020. www.stats.govt.nz/information-releases/gross-domestic -product-march-2017-quarter.

Stewart, A.V., and Charlton, J.F.L. 2006. *Pasture and forage plants for New Zealand*. Grasslands Research and Practice Series No 8., 3rd ed. Palmerston North: New Zealand Grassland Association.

Thom, E.R., Waugh, C.D., Minnee, E.M.K., and Waghorn, G.C. 2006. 'A new generation ryegrass endophyte — the first results from dairy cows fed AR37.' In *Proceedings of the 6th International Symposium on Fungal Endophytes of Grasses,* 13: 293–6. Palmerston North: New Zealand Grassland Association.

Glossary

broadcasting The scattering of seeds randomly over a wide area (often hill country).

continuous stocking (or set stocking) The practice of allowing animals unrestricted access to an area of land for the whole or a substantial part of the grazing season.

dry matter (DM) The part of feed that remains when all its water content is removed (as if it had been completely dried).

high-fertility species Plant species that require high levels of fertility, i.e. a good supply of the macronutrients; nitrogen, phosphorus, potassium and sulphur to persist and yield well, e.g. ryegrass, tall fescue.

legume A plant of the legume family that is able to fix atmospheric nitrogen via root nodules. The common species of legumes found in pastures include white clover, red clover and lotus.

low-fertility species Plant species that only require low levels of fertility to persist, i.e. a reduced supply of the macronutrients; nitrogen, phosphorus, potassium and sulphur, e.g. browntop, sweet vernal.

metabolisable energy Dietary energy reaching the tissues of animals in the form of absorbed nutrients, after accounting for losses in faeces, urine and methane gas. Normally expressed as megajoules (MJ) of metabolisable energy (ME) per unit weight (kg) of food dry matter.

nutritive value A measure of the essential nutrients (carbohydrate, fat, protein, minerals) in a feed.

persistence How long a pasture species lasts in the paddock.

rotational grazing The practice of restricting animals in a set area to graze down the pasture and then shifting animals to a new area to graze, in contrast to set stocking.

stocking rate The number of animals per unit of area, normally reported as *x* animals per hectare.

sward A population of herbaceous plants.

tiller The above-ground new shoots on a grass plant. An important component of a crop's shoot system, as a tiller grows and develops, additional tillers can form in the leaf axils of that tiller. All grasses produce tillers.

The New Zealand Arable Industry

James Millner

Chapter 3

The New Zealand Arable Industry

James Millner

School of Agriculture, Massey University

Introduction

The New Zealand arable industry is primarily a domestic industry, producing cereals for human food and animal feed as well as herbage seeds, processed vegetable crops and forages used as supplementary feed in the livestock industries. Seed of pasture grasses and legumes such as ryegrass and white clover, as well as vegetable seed and frozen vegetables, are exported.

Arable means 'able to be ploughed or cultivated'; land suitable for crop production is generally flat or of gentle, rolling topography with fertile, well-drained soil. Significant areas of arable soil can be found in regions including Waikato, Hawke's Bay, Manawatū, Otago, Southland and, in particular, Canterbury, which has the largest area of arable land in New Zealand. The requirement

for dry weather during the crop harvest season (January–February) means that regions with high summer rainfall (western regions) are generally less suited to arable farming than eastern regions. Wet weather during harvesting can ruin crops and increase costs due to the need to artificially dry grain crops down to a required moisture content, usually 14 per cent, for safe long-term storage. Where summers are typically dry, limiting crop production, many farmers utilise irrigation to minimise the risk of drought.

Many arable farmers in New Zealand are part-time or mixed arable, obtaining a significant part of their income from other farm enterprises including sheep, beef and dairy grazing. Advantages include diversity of income, which means that arable farmers don't have all their eggs in one basket and so reduce risks from bad weather and low product prices. Disadvantages include operating complex production systems and poor economies of scale. The latter is one reason why New Zealand crop producers are unable to compete with overseas growers, for example Australian growers producing commodity cereals such as wheat, barley and maize. Other reasons include high internal transport costs and high machinery costs compared with overseas.

Average yields (production per unit area) in New Zealand are high by global standards (Table 3.1), a reflection of a relatively drought-free climate or the availability of irrigation, fertile soils and skilled growers. Mean yields are increasing by about 1.5–2.0 per cent each year in New Zealand and globally. Crop production systems in New Zealand are relatively intensive, with high use of inputs such as seed, fertiliser and pesticides used to support high-yielding crops.

Table 3.1 Mean yield (tonnes/ha) of wheat in New Zealand, Australia and the United States.

	New Zealand	Australia	USA
Yield	8–9	3–3.5	2–2.5

However, because the New Zealand arable industry is not internationally competitive for many of the important commodity products, such as wheat, much of our arable land is used for livestock production rather than to produce crops. The result of this is loss of self-sufficiency; for example, New Zealand was self-sufficient in wheat up until the 1970s but today most of the wheat we need for staple items such as bread is imported, mostly from Australia or North America (Figure 3.1). Currently New Zealand imports around 1 million tonnes of cereal grains and soybean meal each year. This has implications for

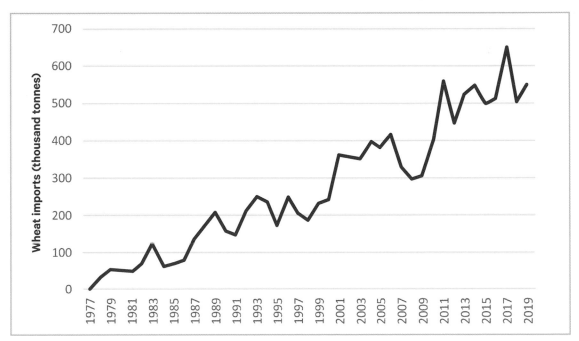

Figure 3.1 New Zealand wheat imports from 1977 to 2019.

food security if natural events such as drought or fire affected production overseas and New Zealand was unable to source its requirements.

Human civilisation depends on a small number of arable crops: the world's three major cereals, wheat, rice and maize, provide almost 70 per cent of global human food intake. Cereals, particularly maize, are also widely used for animal feed, especially for pigs and poultry but in the northern hemisphere this may include intensive beef and dairy cattle production which usually means that animals are housed and fed in feedlots. Barley is also utilised as a feed grain for pigs in many developed countries and for brewing beer, an important activity in many countries.

The ability of the world to feed itself is an important issue. As people's incomes rise, the proportion of cereals in the diet tends to decrease while the consumption of red meat and dairy products increases. This has implications for the environment and the ability to feed ourselves as well as the incidence of so-called lifestyle diseases such as type 2 diabetes and obesity which are rarely found in people living in less developed countries. Many commentators hold that we should reduce our consumption of red meat and dairy products and increase our consumption of complex carbohydrates found in grains, not only to reduce the environmental footprint of agriculture and facilitate healthier diets but also to significantly increase the availability of food on a global scale.

We can get an idea of the magnitude of the potential benefits of this strategy by looking at the number of people able to be fed by a set area of land growing grain crops which can be used for human consumption or used to grow pasture to feed dairy cows. Suppose that 1 ha (100 m × 100 m) of land is used to grow wheat and produces a grain yield of 5000 kg/ha. How many people can be fed by the crop for a year?

The average person requires about 8.4 gigajoules of energy/year if sourced from wheat, allowing for digestibility and losses associated with harvesting, storage, processing, etc. Wheat has an energy content of about 17 gigajoules/tonne, meaning that each tonne of wheat can feed 2 people. A 1 ha paddock of wheat producing a yield of 5 tonnes/ha can therefore feed 10 people for a year.

How many people could be fed if the paddock had been used to grow pasture? Assuming that this paddock produces 12 tonnes of dry matter/ha/year, the output of milk from 1 ha would be about 12,000 litres, with most of the energy in the lactose, fat and protein components. This milk has an energy content of 2750 kilojoules/litre or 33 gigajoules/ha. Using the energy allowance for humans above, this means that 33 gigajoules/8.4 gigajoules = 3.9 people can be fed for a year, well below what can be achieved by production of crops.

Other ways to compare these production systems include the output of protein rather than energy. Analysis of protein outputs from different production systems also shows that crop production on arable soils is able to support greater numbers of people than are livestock production systems. Livestock does have its place, however, such as in areas where topography does not allow the cultivation and production of cereal crops, for example hill country. Under these circumstances, animals able to graze pasture (which is of little nutritional value to humans) are very useful.

The two main categories of arable crop in New Zealand are cash crops (Table 3.2) and forage crops (Table 3.3). Cash crops are produced for sale to generate income directly, whereas forage crops are used to feed livestock which generate income.

Cash crops

A wide variety of different cash crops are grown (Table 3.2). New Zealand is not self-sufficient in the important commodity cereal grains such as wheat, so we import much of our requirement from large overseas producers such as Australia. A key reason for this is that these crops are not sufficiently profitable to compete with more-lucrative crops such as herbage and vegetable seeds or with some livestock systems. This has resulted in a large move away from arable production to dairying in several regions, including Canterbury. Other minor crops, such as oilseed rape (or canola) used to make products such as margarine and, more recently, hemp for human consumption, are also produced on a small scale.

Each crop has its own production, processing and marketing structure. Growers contribute levies to the Foundation for Arable Research, an industry organisation that coordinates and funds research and grower education programmes. Canterbury is the most important region for almost all of the cash crops listed in Table 3.2. For crops such as wheat and barley, 60–80 per cent is typically produced in Canterbury; while for specialised crops such as hybrid carrot seed, 100 per cent is produced there. Exceptions include maize grain, most of which is grown in Waikato; sweetcorn and buttercup squash, mostly grown in Gisborne and Hawke's Bay; and onions, mostly grown around Pukekohe.

Clockwise from top left: wheat, barley, maize, oats.

Table 3.2 Cash crops in different end-use categories produced in New Zealand.

Cereals	
Human consumption	**Animal feed**
Milling wheat	Feed wheat
Malting barley	Feed barley
Oats	Maize
Maize	

Vegetable crops, both fresh and frozen	
Fresh	**Frozen**
Onions	Potatoes
Buttercup squash	Peas
	Sweetcorn

Herbage and vegetable seeds	
Vegetable seeds	**Herbage seeds**
Peas	Ryegrass
Hybrid carrots	White clover
Radish	

Table 3.3 Important forage crops in New Zealand.

Summer crops	Winter crops
Rape, turnips, sorghum — often grown as a drought buffer and grazed (normally break-fed using portable electric fencing)	Brassica species, e.g. turnips, swedes, rape and kale
	Greenfeed cereals, e.g oats
	Italian ryegrass
Whole-crop maize, oats and triticale — these are conserved as silage or baleage (oats and triticale)	Fodderbeet

Forage crops

The purpose of forage crops is the production of animal products such as milk and meat from plants, either by in situ grazing or by harvesting and conservation for later use as supplementary feed. Most livestock in New Zealand is fed a pasture-based diet, but in some seasons pasture growth is often too low to maintain animal production, usually because of low temperatures in winter and low rainfall in summer. During these times, many farmers will utilise forage crops to provide supplementary feed for their livestock.

Forage crops are managed specifically for either summer or winter production, but are also grown for specific markets or clients, with ongoing relationships between grower and purchaser becoming common. A good example of this can be found in Canterbury, where arable farmers may grow forage crops for feeding dairy cows on neighbouring farms over the winter.

The area planted in forage crops each year is greater than the cash crop area. The most important forage crops are kale, swedes and fodder beet (winter feed), as well as turnips (summer feed) and maize (silage). The area planted in maize for silage is more than twice that planted for grain.

Principles of crop production

The basis of crop production is the utilisation of plants (crops) to intercept and convert solar radiation (light energy) into food (chemical energy) through photosynthesis. The general relationship between crop yield and solar radiation interception is shown in Figure 3.2. Yield is the product of the total amount of radiation

Clockwise from top left: kale, maize, greenfeed oats, swedes.

intercepted by the crop and the efficiency of its conversion to dry matter, represented by the slope of the line.

The main objective for crop managers wanting to maximise yield is to maximise total radiation interception. The key factors determining radiation interception are leaf area (expressed in m^2 of leaf per m^2 of ground, or *leaf area index*) and the duration of interception (Figure 3.3).

Leaf area index

Many crops require a leaf area index of 3 to 4 to intercept the majority of incoming radiation. Warm temperatures increase the rate of canopy development whereas cold temperatures reduce development. Drought can also reduce leaf area index through wilting or rolling of leaves. Examples of the methods farmers can use to influence leaf area include the use of irrigation and herbicides and the application of nitrogen fertiliser.

Leaf area duration

The longer that crops are able to intercept and utilise radiation, the more biomass they will accumulate. This means that early maturing crops usually yield less than late maturing crops, and

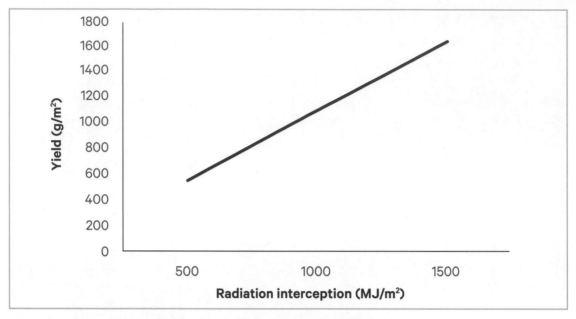

Figure 3.2 The relationship between dry matter yield and intercepted sunlight for a wheat crop. The radiation use efficiency (slope) is 1.1 g of dry matter (DM)/Megajoule (MJ) of intercepted sunlight.

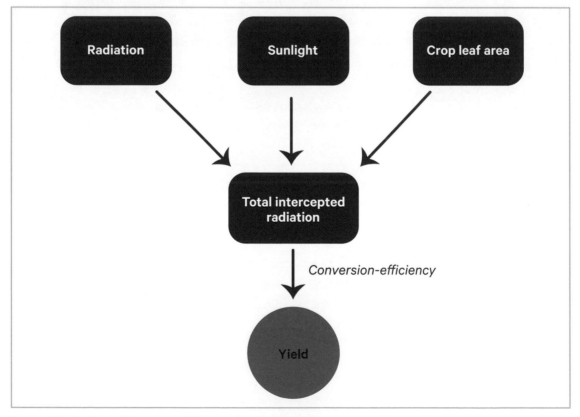

Figure 3.3 A simple model for crop yield development.

crops that prematurely die off due to drought or disease will suffer reduced yield.

Harvest index

Many of the crops important for human consumption are not completely utilised. Only the seed or fruit is eaten because they contain carbohydrates and proteins that humans are able to digest. The rest of the crop, including the stem and leaves, are fibrous (containing a lot of cellulose) and cannot be digested by human digestive systems. Ruminants, such as sheep and cattle, can digest fibre. The proportion of the total biomass present in the harvested component is known as the *harvest index*. It is stable across a wide range of crop yields, which provides farmers with a clear objective — grow more total biomass to achieve higher grain yields.

Humans began growing crops such as wheat around 10,000 years ago in the Middle East and have been selecting plants which produce higher yields ever since. Most of the genetic improvement for grain yield in crops such as wheat and rice has been based on selecting plants which are shorter but have a higher percentage of the total biomass in the seed or grain. The ratio of grain yield to total yield gives the harvest index. Modern cereal crop varieties have a high harvest index compared with old varieties.

Potential crop yield

Potential crop yield is determined by the environment in which the crop is growing. Sites with fertile soils, good rainfall and ideal temperatures have high potential crop yields. At sites where one or more of these factors is limited, yield potential is reduced. The reason for high crop yields in New Zealand compared with most other countries (see Table 3.1) is due to a climate with moderate temperatures and either good rainfall or the availability of irrigation where rainfall is limited.

Crops with good canopy cover (high leaf area index) will intercept 95 per cent or more of incoming sunlight. This sunlight (radiant energy) is converted into biomass (chemical energy) at a rate of about 1 gram (g) of dry biomass per megajoule (MJ) of intercepted sunlight, but this can vary so if the crop is under drought stress then the conversion rate will be lower. This is because under drought stress, the stomata on the leaves will close to prevent moisture loss and this also prevents uptake of CO_2 needed for photosynthesis.

This information can be used to estimate how fast crops are growing. The other information required is the amount of energy available in sunlight each day. In Palmerston North, for example, about 22 MJ/m² of energy arrives every day during December. A crop intercepting 95 per cent of this sunlight and converting it to biomass at a rate of 1 g/MJ of intercepted sunlight is growing at 22 MJ/m²/day × 0.95 × 1 g/MJ = 20.9 g/m²/day. This yield is best converted to the units typically used in agriculture, kilograms/hectare/day. One hectare (ha) = 10,000 m² and 1 kilogram (kg) is 1000 g — so multiply by 10,000 and divide by 1000 (i.e. multiply by 10) to convert g/m² into kg/ha. The net result is a growth rate of 209 kg/ha/day of dry matter. The longer the crop is able to keep intercepting radiation, the higher the final yield will be (see Figure 3.3).

The major reason for differences in crop yield between sites, seasons and management is the amount of leaf area produced and retained by the crop. Examples of management influences on radiation interception include sowing date, control of weeds/pests/diseases and fertiliser application. Examples of major environmental influences on crop yields include drought. Drought reduces leaf area, growth duration and the efficiency of conversion of sunlight to biomass, hence the emphasis on irrigation in regions where summer rainfall is limited. A typical symptom of drought in plants is leaf wilting. This immediately reduces leaf area and eventually growth duration.

Composition of cereals

Cereal grains are an ideal feed source for both humans and animals (particularly monogastric species such as pigs) because the carbohydrates and proteins they are made up of are readily digested. In regions of the world with low incomes, the diets of many people are dominated by cereal grains such as wheat, rice and maize. However, diets based on these grains are deficient in protein (Table 3.4), so legume crops such as soya beans (28 per cent protein) are important to alleviate protein deficiency if consumption of meat or fish is minimal. Animal feeds based on cereal grains also need to be fortified with protein for adequate nutrition.

Table 3.4 Approximate composition (%) of wheat grain and white wheat flour.

	Grain	White flour
Carbohydrate	71	85
Protein	12.5	10
Fibre	13	4
Fat	1.5	–
Other	–	1

Much of the protein in wheat comprises gluten, which is essential for producing consumer products such as bread and buns but causes coeliac disease in a small percentage of people, due to an intolerance of gluten. Table 3.4 highlights the nutritional value of wholegrain flour versus refined white flour. Wholegrain flour has the same composition as the grain whereas white flour has most of the fibre removed, increasing the percentage of carbohydrates. The use of white flour represents a problem for diets in many industrialised countries — high intake of processed foods with insufficient dietary fibre is associated with health problems such as cancer, obesity, diabetes and heart disease.

Management on arable farms

Management on arable farms includes the short-term management associated with establishing, growing and harvesting different crops as well as longer-term management associated with selecting which crops to grow in different paddocks over time. Most farmers avoid growing the same crop successively in the same paddock for too long because of the build-up of pests and diseases in the soil, which increases costs and reduces yields. Crop rotation is usually adopted to help prevent this.

The annual cycle outlined in the box is a typical example from a mixed arable farming system. Variations occur according to local conditions, particularly climate and availability of markets. This schedule is based on spring-sown grain crops.

A typical annual cycle: Mixed arable farming system

Winter (June–August)

The year begins at this time with planning and economic analysis of different cropping options. In some regions with low rainfall or with light soils, cultivation may begin in late winter.

Spring (September–November)

Most spring-sown crops are planted in August and September; soil cultivation often begins several weeks prior to planting so that the seedbed is ready for planting at the desired time. Wet soil conditions can delay cultivation and consequently planting, which can reduce yields. Cultivation usually begins with ploughing, which serves two purposes. It buries existing vegetation, killing it, and it also results in increased soil temperatures. This stimulates greater microbial activity which in turn results in the release of plant nutrients such as nitrogen and phosphate from decaying organic matter.

Establishment begins with drilling (sowing) seed, usually together with fertiliser (nitrogen and phosphate). Drilling into a cultivated seedbed is common in New Zealand but many crops are planted using direct drilling techniques. This involves using a drill capable of placing seed into uncultivated ground which may have a lot of dead plant material on the soil surface. Because existing vegetation has not been controlled by ploughing, direct drilling usually requires the use of herbicide to control the existing vegetation. The most commonly used herbicide for this purpose is glyphosate (see note 1). The benefits of direct drilling include retention of soil organic matter and less soil erosion from rain and wind.

The seeds used are typically the result of breeding programmes designed to select varieties with a range of traits including high yield, resistance to pests and diseases, resistance to drought, and good grain quality. Grain quality can include physical properties such as size, and nutritional properties such as protein content. Genetic improvement for desirable traits is ongoing and has resulted in the yields of many crops increasing by 1–2 per cent globally for many decades; this has been crucial to the maintenance of food security and political stability worldwide. Plant breeders are using new technologies to increase the productivity of many crop species, including controversial methods such as genetic modification (see note 2). Future challenges include an increasing global population and the effects of climate change leading to loss of arable land to rising sea levels and increased weather extremes.

Seed quality is important; most crop seed is tested for its ability to germinate and for the presence of weeds and other impurities before planting.

Crops tolerant of low temperatures are planted earlier in spring, but crops requiring warm temperatures for growth (e.g. maize) are planted later in spring to ensure that soil temperatures are sufficient for germination. Some cereal crops may be planted in the autumn. The advantage is higher yield, but the disadvantage is that only one crop per year is possible.

Shortly after crop emergence, weeds are controlled, usually with the use of herbicides. Weeds reduce crop yields by competing for light, moisture and nutrients.

Crops are monitored as they develop for the presence of insect pests and fungal diseases that may reduce yield and/or crop quality. Insecticides and fungicides may be applied if control is warranted. Most pests and diseases are wind-borne; that is, they spread from crop to crop through the action of wind. Pests spread this way include aphids; aphids often spread viral diseases such as cereal yellow dwarf virus. Many fungal diseases are spread by air-borne spores; potentially damaging examples in New Zealand for wheat include leaf rust, stem rust and speckled leaf blotch, and for barley, net blotch. Soil-borne diseases are described in the section on crop rotation (see page 68).

Additional fertiliser, usually nitrogen, may also be applied if required. This may be used not only to increase yield but also to increase the protein content in wheat crops destined for bread-making. High-protein flour (a high gluten level) results in greater loaf volume, better texture and better keeping ability of bread.

Summer (December–February)

Early summer is usually a quiet time, as most crops are nearing maturity and don't require many inputs or intervention. By mid-summer the early maturing crops are ready for harvest. Harvest continues through to early autumn for late-maturing crops. Crops are harvested using a combine harvester which cuts, gathers and threshes the crop in one operation, separating the grain from the straw or chaff. Wet weather can delay harvest in some years.

In New Zealand, grain needs to be at 14 per cent moisture content or less for safe long-term storage. Above this level it is vulnerable to degradation from fungi and grain-storage insect pests such as granary weevil. Some farmers may harvest grain at higher moistures and artificially dry it to 14 per cent; however, this entails an extra cost.

Virtually all crops are harvested at this time. An important exception is maize, which under New Zealand conditions doesn't mature until mid to late autumn and must be artificially dried to 14 per cent moisture content.

Autumn (March–May)

Soon after harvest, paddocks are cultivated or direct drilled so that they can be used again quickly. This is usually preceded by removal of the residue (straw) from the previous crop, often by burning, which is quick, inexpensive and effective but risky (fire spreads) and polluting. Occasionally straw is baled and sold.

If the paddock is to be returned to permanent pasture it will be sown with a grass/clover seed mix. However, if it is intended to plant the paddock with another grain crop in the following spring, it may be planted with a temporary forage crop such as greenfeed oats or Italian ryegrass to provide forage over the late autumn–winter period.

Notes

1. Originally this chemical was sold under the brand name 'Roundup'. Its use is controversial in some countries because of claims that exposure to it causes cancer in humans. There is currently very little scientific evidence to support these claims.
2. Genetic modification often involves the transfer of useful genes from one species to another. For example, a gene taken from a common species of soil bacteria, *Bacillus thuringiensis* (the so-called BT gene), transferred into crop species such as maize provides resistance to insect pests.

A high-yielding wheat crop. If this crop has 600 ears/m² (6 million ears/ha) and each ear contains 40 grains weighing 40 mg (1000 grains weigh 40 g) each, then the yield will be 9600 kg/ha.

Crop rotation

Crops are nearly always grown in the same pad-dock for more than 1 season (year). Crops may be grown in a paddock for many years before it is returned to pasture. The sequencing of these crops is often pre-determined, influenced by the availability of markets for different crops, crop fertility requirements, susceptibility of different crops to the build-up of soil-borne pest and disease problems, soil structure and many other specific factors such as availability of labour and preference for different crops. Although the sequence of annual crops in a rotation is pre-de-termined, in practice rotations change, often in response to prices.

There are many benefits associated with crop rotation, outlined below.

Take-all disease of wheat.

Preventing build-up of soil-borne fungal disease

Many fungal diseases build up in the soil with successive susceptible crops. By breaking a sequence of susceptible crops with a resistant crop, many important soil-borne diseases can be controlled. A good example of this is 'take-all' in wheat, caused by a fungus (*Gaeumannomyces graminis*) which builds up quickly under succes-sive wheat crops resulting in large yield losses (Table 3.5). Take-all causes decay of the roots and stem base, and the only viable way to control it is to not grow successive wheat in the same paddock for more than two years. Crops such as peas can be used to break the cycle because the take-all fungus is unable to thrive on legumes.

Utilising different crop fertility requirements

Paddocks that have been in long-term grazed pasture usually have good soil fertility and good soil structure. Cultivation and crop production generally result in reduced fertility, particularly nitrogen content, and poorer soil structure asso-ciated with limited return of nutrients through deposition of dung and urine from grazing ani-mals along with reduction in soil organic matter (carbon), which is crucial for soil structure and

Table 3.5 Effect of successive wheat crops on grain yield and soil. Previously the paddock was in pasture.

Year 1	Year 2	Year 3
8.0 t/ha	7.3 t/ha	3.9 t/ha
Over the 3 years:		
• Soil carbon content declined from 39 to 33.5 t/ha		
• Earthworm populations declined from 750 to 200 worms/m^2		

Source: Francis, Tabley, and White 2001.

fertility. Soil fertility declines with each successive crop (Table 3.6). Consequently, crops which require high fertility are usually grown at the beginning of a rotation (1st year out of pasture) while crops more tolerant of lower fertility are grown near the end of a crop sequence.

Utilising pasture to restore fertility and soil structure

At the end of a crop rotation, paddocks are normally returned to pasture. The length of the pasture phase may be as little as 2–3 years or as long as 10–15 years depending on the proportion of the farm under cultivation each season. The benefits of pasture include restoration of fertility and soil structure and a reduction in many soil-borne fungal diseases.

Cropping systems utilising rotations that include a pasture phase tend to be more sustainable than systems that rely on continuous cultivation and few different crops. Rotation minimises financial risk by diversifying income — minimising exposure to any single market. Seasonal workload can be spread, allowing better utilisation of machinery and labour. Crop yields are higher while crop inputs (e.g. fertiliser and cultivation required to produce a seedbed) are lower, increasing profitability.

Table 3.6 Effect of 10 years of cultivation on soil organic matter and nitrogen content in the top 5 cm.

	Permanent pasture	Cultivated
Organic matter (%)	4.03	2.34
Nitrogen (%)	0.28	0.15

Source: Horne, Ross, and Hughes 1992

References

Foundation for Arable Research. n.d. Accessed 17 February 2020. www.far.org.nz/.

Francis, G.S., Tabley, F.J., and White K.M. 2001. 'Soil degradation under cropping and its influence on wheat yield on a weakly structured New Zealand silt loam.' *Australian Journal of Soil Research* 39 (2): 291–305.

Horne, D.J., Ross, C.W., and Hughes, K.A. 1992. 'Ten years of a maize/oats rotation under three tillage systems on a silt loam in New Zealand. 1. A comparison of some soil properties.' *Soil and Tillage Research* 22 (1–2): 131–43.

New Zealand Flour Millers Association. n.d. Accessed 17 February 2020. www.flourinfo.co.nz/.

United Wheatgrowers. n.d. Accessed 17 February 2020. www.uwg.co.nz/news.cfm.

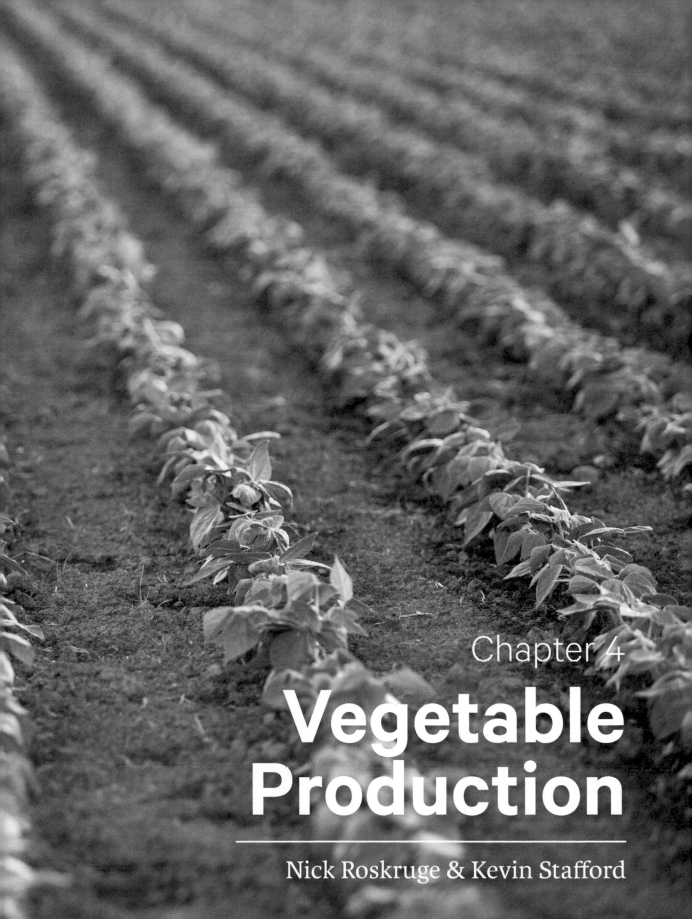

Chapter 4

Vegetable Production

Nick Roskruge & Kevin Stafford

Chapter 4

Vegetable Production

Nick Roskruge and Kevin Stafford

School of Agriculture and Environment, Massey University

Introduction

New Zealand's horticultural industry is based on the domestic markets for the fruit and vegetables produced (Figure 4.1; Table 4.1). In 2019, New Zealand households spent an estimated $890 million on fresh and chilled vegetables and $390 million on processed vegetables (all values in New Zealand dollars). Increased incomes have changed eating habits over the past 25 years and this has seen an increase in the variety of fruit and vegetables grown and consumed. While there has been minimal change in the global consumption of food categories such as cereals, pulses and oilcrops over this period, the fruit and vegetable category has flourished with global consumption rising from approximately 175 kg per capita per year in the early 1990s to well over 250 kg per capita per year in recent years.

In 2019, New Zealand exported vegetables worth $699 million (2018: $625 million) of more than 20 significant types to many destinations (Figure 4.3). The average export value for the five years prior to 2018 was $607.4 million. The biggest export earner is onions. Onion export values have seen the largest increase in value over the past few years. Onions together with potatoes, squash, peas and vegetable seeds were among the top 10 exports of New Zealand's total horticultural production in 2019.

New Zealand's reputation as a supplier of top-quality produce means that the industry should look to target increased exports into developed markets in countries such as the United States, Japan, Australia, the European Union and Canada where consumers could pay premium prices (Figures 4.2 and 4.3). New Zealand succeeds in the fruit and vegetable export arena due to factors such as strong industry organisation, an ability to supply counter-seasonal produce, and the ease of doing business with Kiwi companies. Another market where New Zealand targets increased exports is the Pacific region.

Māori were the first to grow vegetables commercially. They traded with early settlers, and exported potatoes to Sydney. As European migrants took up land and cleared forests, European market gardening took over.

Land suitable for vegetable production is found throughout New Zealand. Potatoes have been grown here since the late eighteenth century. Until the 1960s, they were bought fresh and cooked at home; now, more than half the potatoes grown are turned into French fries or crisps. Half a million tonnes of potatoes are grown annually and most are sold locally. Onions are grown mainly in Waikato, Auckland and Hawke's Bay. The climate and soils of Gisborne and Hawke's Bay are particularly good for growing squash.

A look at New Zealand's overall vegetable industry reveals the wide diversity of crops produced and the diverse uses these crops can be put to. We have become used to seeing fresh vegetables available all year round, and sometimes products are imported during our off-season to meet market demands.

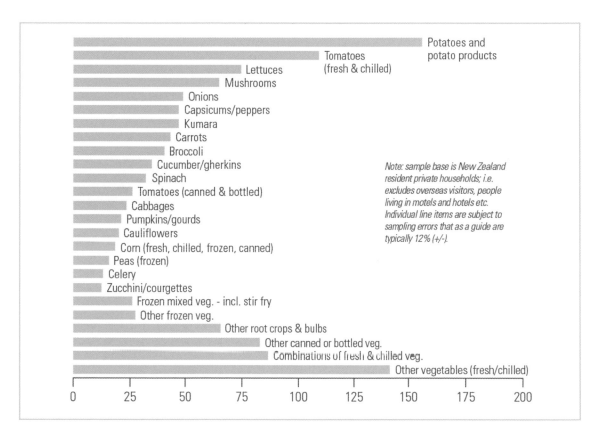

Figure 4.1 Consumer spending on vegetables based on New Zealand resident private households (2019, $ million). (Source: Fresh Facts 2019)

Table 4.1 Vegetable production and sales values for 2018.

Fresh and processed vegetables				Sales value ($ million)		
	Growers[c] (no.)	Planted[c] area (ha)	Crop volume[c] (tonnes)	Domestic[c] 2018	Exports 2018[g]	
					Fresh	Processed[b]
Asparagus	42	570[c]	1800	8.5	0.7	
Beans	30	1200[c]	24,700	10.3		
-fresh	5	300	3000	6.0		39.8
-processed	25	900	21,700	4.3		
Beetroot	28	380	27,500	14.8		
-fresh	20	120	6000	4.0		
-processed	8	260	21,500	10.8		25.0
Brassicas	125[e]	3432[c]	115,700	80.3		
-Broccoli	75	2082[c]	24,700	35.0		
-Cabbage	75	804[c]	58,000	25.3	2.0	
-Cauliflower	20	546[c]	33,000	20.0		
Capsicums	22	85[c]	21,000	25.0	21.0	
Carrots	54	1410[c]	153,900	56.0		
-fresh	20	800	88,000	40.0	8.5	
-processed	34	610	65,900	16.0		1.8
Cucumbers[h]	51	71[c]	2000	20.0		
Eggplant/ Aubergines	20		1000	8.5		
Garlic	10	210[c]	1200	7.0	0.8	
Kumara[h]	48	2541[c]	24,000	55.0		
Lettuces	162	1582		42.0		
-outdoor	140	1532[c]		17.0	0.6	
-greenhouse	22	50[c]		25.0		
Melons	20	273[f]	4800	28.0	1.3	
Mushrooms[h]	7	25[c]	8500	42.0	1.4	
Onions	92	5227[c]	191,639	30.0	93.0	

Fresh and processed vegetables			Sales value ($ million)		
	Growers[c] (no.)	Planted[c] area (ha)	Crop volume[c] (tonnes)	Domestic[c] 2018	Exports 2018[g]
					Fresh	Processed[b]
Potatoes			527,190	139.0		
-fresh/table	171	10,344[c]	150,788	56.0	26.4	114.9
-processed			354,360	83.0		
Pumpkins	30	1158[c]	38,000	13.0		
Shallots	4	30[c]	1200	3.0	0.2	
Silverbeet/Spinach	10	2082[f]	3500	12.0		
Squash	24	6642[c]	88,179	3.0	58.6	
Sweetcorn			98,800	27.5		
-fresh	179	3871	22,000	11.0	0.1	42.0
-processed			76,800	16.5		
Tomatoes	129	528	95,400	141.5		
-outdoor, processed	6	408[c]	53,000	8.5		3.1
-greenhouse	123	120[c]	42,400	200.0	9.6	
Truffles[d]	75	70	0.2	0.5		
Mixed vegetables	Made from combinations of the above crops					25.3
Dried vegetables	Excluding peas, beans, corn					11.3
Vegetable preps						3.3
Vegetable juices						31.7
-carrot juice[d]						28.0
-other veg. juices						3.7
Other vegetables[a]					2.2	10.5
Total	800[e]	45,466[f]			226.4	396.4

Crops areas are predominantly sector estimates. [a] Includes taro, celery, parsnips, spring onions, Asian vegetables (excl. Chinese cabbage), yams, witloof, leeks, vegetable shoots, shallots, swedes and some others. [b] Processing includes freezing, canning, juicing and artificial drying. [c] Sector estimates. Blank entries indicate that the information is not available. [d] Authors' estimates. [e] Growers produce multiple crops. [f] Statistics New Zealand Production Census crop areas as at 30 June 2017. [g] Statistics New Zealand from export entries. [h] Crop grown both outdoor and indoor/protected. (Source: Fresh Facts 2018)

Within the horticulture industry there are a number of sectors, each represented by one or more interest groups. These include:

- Horticulture New Zealand (www.hortnz.co.nz) — This organisation represents the interests of much of New Zealand's horticultural producers and provides considerable resources and support through a range of initiatives. It is also responsible for the periodical publication *Grower*, a journal that focuses on the vegetable industry.
- Tomatoes NZ (www.tomatoesnz.co.nz/) — This fresh tomato group has established a web presence to provide a range of information for fresh tomato growers and a timely resource regarding industry events and developments for the general public. The website is funded from the commodity levy paid by New Zealand's commercial fresh tomato growers. The group represents around 155 growers across the country who collectively grow around 102,900 tonnes of tomatoes annually.
- Potatoes NZ (www.potatoesnz.co.nz) — This group hosts an online library on its website with relevant publications (year reports and the like). The group represents around 170 growers across the country, who collectively grow over half a million tonnes of potatoes.
- National Māori Vegetable Growers Collective (www.tahuriwhenua.org) — This collective, known as Tahuri Whenua Inc. Soc., was established to represent Māori interests in the national vegetable sector. It provides a conduit for grower interactions, extension and research and development opportunities. The group holds regular meetings around the country and provides a basis for industry extension to Māori land-owners.

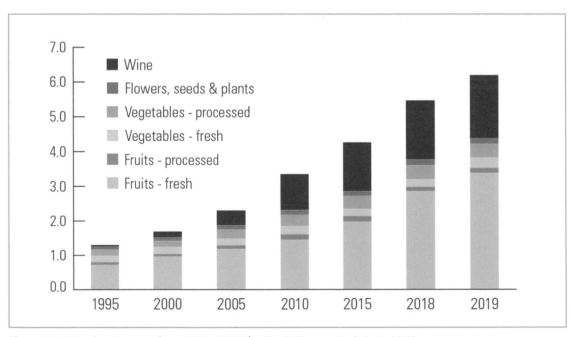

Figure 4.2 Horticultural exports from 1995 to 2019 ($ million). (Source: Fresh Facts 2019)

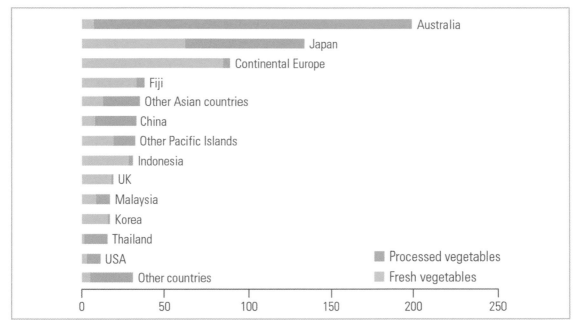

Figure 4.3 Destinations of vegetable exports ($ million, 2019). (Source: Fresh Facts 2019)

Principles of vegetable production

Vegetables represent a diverse collection of plants and plant parts. The common vegetables grown in horticultural commercial systems in New Zealand are listed in Table 4.2.

Vegetables are produced under a range of production systems that can be defined as follows:

- **Annual crops, short-term** — e.g. lettuce, brassica; those with short production cycles and capable of more than one crop per 12-month period.
- **Annual crops, long-term** — e.g. potatoes, onions; only one crop cycle possible in a year.
- **Perennial crops** — e.g. asparagus, globe artichoke; can be harvested annually for several years, e.g. asparagus has the ability to crop for up to 20 years.

- **Indoor crops** — e.g. tomato, cucumber; can be annual short-term such as cucumbers, annual long-term such as tomatoes, or perennial such as strawberries.
- **Outdoor crops** — e.g. leeks, yams; generally paddock-grown but can include extensive crops such as onions and potatoes, intensive crops such as parsley and anything in between.
- **Herbs** — e.g. dill, parsley; can apply to a very diverse range of herbaceous plants.
- **Niche** — e.g. sprouted beans, mushrooms; very specialised systems established.
- **Organic** — can apply to all crops and is often philosophically driven; different organic systems and standards are accessible by growers in New Zealand. e.g. BioGro, Biodynamics.
- **Nursery and seedling specialists** — applies to all crops above.

Following are some of the types of choice that growers have available to them when planting.

Table 4.2 Common vegetables and fruits in New Zealand, and their markets.

Vegetable	Latin name	Markets			
		Fresh	Processed	Domestic	Export
Onions	*Allium cepa*	Y			Y
Potatoes	*Solanum tuberosum*	Y	Y		Y
Squash (kabocha)	*Cucurbita maxima*	Y		Y	Y
Tomatoes	*Solanum lycopersicum*	Y	Y	Y	
Asparagus	*Asparagus officinalis*	Y	Y		Y
Peppers (capsicums)	*Capsicum annum*	Y	Y	Y	
Lettuce	*Lactuca sativa*	Y		Y	
Cauliflower	*Brassica oleracea* var. *botrytis*	Y	Y	Y	
Sweetcorn	*Zea mays* var. *saccharata*	Y	Y	Y	
Broccoli	*Brassica oleracea* var. *italica*	Y	Y	Y	
Cucumbers	*Cucumis sativas*	Y		Y	
Kūmara (sweet potato)	*Ipomoea batatas*	Y		Y	
Yams	*Oxalis tuberosum*	Y		Y	
Cabbage	*Brassica oleracea* var. *capitata*	Y	Y	Y	
Garlic	*Allium sativum*	Y		Y	
Pumpkin	*Cucurbita pepo*	Y		Y	
Peas	*Pisum sativum*		Y	Y	Y
Beans	*Phaseolus spp.*		Y	Y	
Carrots	*Daucus carota* var. *sativus*	Y	Y		Y
Brussels sprouts	*Brassica oleracea* var. *gemmifera*	Y		Y	
Leeks	*Allium porrum*	Y		Y	
Parsley	*Petroselium crispum*	Y		Y	
Beetroot	*Beta vulgaris* var. *rubra*	Y	Y		Y
Spinach	*Spinacia oleracea*	Y		Y	
Rhubarb	*Rheum rhabarbarum*	Y		Y	
Spring onions	*Allium fistulosum*	Y		Y	
Mushrooms	*Agaricus bisporus*	Y*	Y	Y	Y
Sprouted beans & seeds	(various)	Y		Y	
Parsnips	*Pastinaca sativa*	Y	Y	Y	
Celery	*Apium graveolens* var. *dulce*	Y		Y	
Eggplant (aubergine)	*Solanum melongena*	Y		Y	
Radish	*Raphanus sativus*	Y		Y	
Watermelon	*Citrullis lanatus*	Y		Y	

* And dried.

- **In-situ planting**, also known as direct seeding/planting — seeds are sown directly in the space they will grow in through to maturity.
- **Bare-root transplants** — transplants produced in nursery systems are planted directly into the ground or indoor production system where they will be grown to maturity.
- **Cell transplants** — plants produced in specialist systems; these transplants are generally more expensive to produce.
- **Fresh market** — markets are generally local, requiring good post-harvest systems.
- **Processed market** — adding value to produce and extending its marketability.
- **Harvest** — all year (sequential) versus part-year (often a one-off harvest).
- Plants with specific **harvest 'windows'** — e.g. asparagus, only harvested in spring.
- **Commodity crops** — generally grown in bulk volume because they may be of lower value by weight.
- **Niche market crops** — high value and often export-oriented.

Many crops are grown based on their suitability to specific regions, e.g. carrots and Brussels sprouts are grown in the Ohakune region. However, the development of specialist growing systems means that some crops can be manipulated and grown almost anywhere, any time. An example of this is the out-of-season production of asparagus in covered systems.

Basic plant physiology determines the management of the crops pre- and post-harvest. For example, lettuce is harvested as immature leaves with a large surface area; these leaves can dehydrate quickly because of the immature nature of their cells and because of the leaf area, so post-harvest management takes these points into account. A potato is a tuber, a modified

Important external factors in producing vegetables

- Soil type (and history)
- Soil drainage
- Climate — advantages
- Climate — limitations
- Nutrient levels — management
- Fertilisers/manures
- Crop seed choice
- Varieties — match to resources and demands
- Viability
- Method of propagation
- Weeds — management
- Shelter
- Pest populations and issues
- Disease issues
- Available budget
- Access to expertise
- Support industries, e.g. cool chain or transport

storage organ sourced from a mature plant, and post-harvest management of potato tubers will differ greatly from that undertaken for lettuce. The most recognisable vegetables and the factors of interest to production and post-harvest management are shown in Table 4.3.

Producing vegetables successfully requires an understanding of a wide range of factors. Some are generic to any crop, while others are specific to a particular crop or region. In all cases the knowledge around these factors creates a matrix that can then be used as part of any decision-making regarding production.

Table 4.3 The harvestable parts of common vegetables.

Plant part	Vegetable examples	Notes
Fruits	Tomato	Essentially fruits utilised as vegetables
	Aubergine/eggplant	
	Pumpkin	
	Zucchini/courgette	Most cucurbits are harvested as fruits at various stages of maturity
Seeds	Beans	Seeds contained within pods and eaten at an immature stage
	Peas	
	Sweetcorn	Harvested immature for markets
Leaves	Lettuce	Immature plants, subject to water loss because of the large leaf area
	Silverbeet	
	Radicchio	
Flowers	Broccoli	Immature (multiple) flower heads
	Globe artichoke	Flowers harvested singularly
Bulbs	Onions	Specialised storage organs for carbohydrates used to initiate the next generation of the plant
	Shallots	
Roots	Carrots	
	Kūmara	
Tubers	Potatoes	
	Yams	
Stems	Rhubarb	Perennial with large taproots
	Asparagus	Perennial with dormancy over winter
Fungi	Mushrooms	Flowering bodies
Whole plant	Spinach	Single harvest of whole immature plants
	Kōkihi New Zealand spinach	

Vegetable crops come from a range of plant families. The relationship between plants within the same family grouping is important when deciding on production systems or considering weed or pest and disease issues.

Plant pests and diseases are often aligned to families of plants (Table 4.4), and growing crops continuously from the same family can therefore create issues with threshold numbers of pathogens being carried between crops. Additionally, in crop rotation systems the different crops will utilise different profiles in the soil or atmosphere. They can therefore complement each other in a rotation, as they do not tax the same profile continuously. For example, a deep-rooted crop such as potatoes will utilise a different soil profile than a shallow-rooted one such as lettuce. Similarly, a leafy crop will have a higher demand for certain nutrients such as nitrogen over their production cycle than a root crop might have.

The relationship between crops/plants is a factor in the decisions a grower needs to make as a horticulturist. As an example, all plants belonging to the Brassicaceae family will have similar soil requirements (pH, nutrients, etc.) and will also suffer from the same pest and disease issues. Another example is the weed management required for sweetcorn — as this belongs to the grass family, it may therefore have to compete with other weed grasses during establishment.

The list of plants by family given in Table 4.4 identifies some of the key crops and their relationship to other crops.

Table 4.4 Vegetable crop families.

Family	Examples
Apiaceae	Carrot, celery, parsnip, parsley
Asteraceae	Daisies: globe artichoke, salsify, lettuce, chicory, endive
Brassicaceae	Cauliflower, cabbage, broccoli, Brussels sprouts, kale, kohlrabi, Chinese cabbage, swedes, turnips, mustard, watercress, radishes
Chenopodiaceae	Silverbeet, spinach, beetroot
Convolvulaceae	Kūmara (sweet potato)
Cucurbitaceae	Pumpkin, squash, gherkin, cucumber, zucchini (courgette), melon, kamokamo
Fabaceae	Legumes: peas, beans, lentils, soybeans
Liliaceae	Asparagus, garlic, leeks, onions
Poaceae	Grasses: sweetcorn
Polygonaceae	Rhubarb, sorrel
Solanaceae	Potatoes, tomatoes, eggplant (aubergine), tobacco, capsicum

Indoor production

In horticulture, production systems range from being broad and extensive to contained and relatively intensive. Indoor cropping systems represent the more intensive approaches to crop production and allow the grower to have greater control of the environment at all stages of crop production.

Indoor systems target a maximum use of space relative to yield and returns. They also allow managers to control inputs such as temperature, water, nutrients, light, gaseous levels and exchange (for example, CO_2). Several systems are utilised indoors, the most common being the hydroponic or NFT (nutrient film technique), followed by drip irrigation systems where the plants are grown in specialist bags or containers and the water and nutrients are metered to them throughout the day and night.

The key factors that make indoor systems both valuable and resilient to external factors are:

- potential yields and returns per area
- out-of-season production
- niche markets, e.g. melons for Japan
- temperature controls, including day–night differentials
- light levels, duration and quality
- water — timing, volume, placement and quality
- nutrients — levels, timing, placement
- protection from climatic extremes, e.g. winds
- gas levels, including gaseous exchange in closed systems
- containment of plant health issues — improved hygiene practices
- implementation of IPM (integrated pest management) systems.

Some of the factors that make indoor systems difficult for growers include:

- the investment required for establishment of the system
- the specialist skills required
- the need to keep abreast of technical needs
- the potential for plant pathogens to become entrenched in the system
- higher risk for crop losses over the whole system.

The decision to produce indoor crops is not an easy one and requires careful thought. Once established, though, indoor crops are generally viable and in strong demand from consumers.

The main vegetable crops produced indoors in New Zealand are:

- tomatoes
- peppers
- cucumbers
- lettuce
- watercress
- melons — some high-value varieties
- mushrooms — several varieties, including shiitake and common mushrooms
- sprouted beans and seeds.

Vegetables suited to indoor production

Tomatoes (*Solanum lycopersicum*)

Tomatoes are a fruit that we use as a vegetable. Originating in South America, they were introduced to Europe during the Spanish colonisation of that continent. The tomato is one of the most popular vegetables in New Zealand, second only to potatoes in household spending on vegetables. There are many varieties of tomatoes grown for the fresh market, including the smaller cherry tomatoes. Indoor tomato systems allow for continuous harvests from the same plant for a period of up to 2 years. Tomatoes grown for the processed market are produced outdoors.

The main types of tomato are:

- indoor cultivars suited to glasshouses, truss harvests
- beefsteak tomatoes
- acid-free tomatoes
- cherry or cocktail tomatoes
- outdoor cultivars (seasonal).

Peppers (*Capsicum annum*)

Peppers belong to the Solanaceae family and originate from Central and South America. There are many varieties available, with a range of flavours, shapes and colours. The main pepper varieties grown in New Zealand are:

- sweet or bell peppers (round shape and mild flavours)
- sweet long peppers (long shape and mild flavours)
- miniature bell peppers
- long hot-type peppers — including cayenne, chilli and jalapeño.

Outdoor pepper transplants are planted from October onwards and take about 75 days from transplanting to harvest (around 160 days from sowing). Indoor production goes on all year. Peppers are sequentially harvested in both systems and require good cool storage post-harvest.

Cucumbers (*Cucumis sativus*)

Cucumbers are a popular salad ingredient, and several varieties are grown indoors in New Zealand. The main ones are:

- telegraph cucumber — the most popular variety
- short cucumber
- Lebanese cucumber (a recent introduction to New Zealand)
- apple cucumber.

The cucumber fruit are harvested relatively immature and therefore require good post-harvest management. They can be grown all year round although demand tends to be seasonal.

Lettuce (*Lactuca sativa*)

Lettuces are one of our most popular vegetables and we utilise them in many ways. In recent years the introduction of loose-leaf, mixed lettuce packs has broadened our use of this leafy vegetable. The majority of lettuces produced are the standard lettuce sometimes known as an iceberg variety or crisphead. Lettuces are generally grown in NFT or hydroponic systems. The main lettuce types are:

- iceberg (crisphead)
- butterhead (buttercrunch — a loose-leaf variety)
- cos — a winter lettuce also known as romaine lettuce
- endive (*Cichorium endive*)
- loose-leaf lettuces such as Red Sails, Green Oak, Red Oak, Lollo Rossa.

Melons (*Cucumis melo*)

Some high-value markets exist for certain melon varieties. Melons belong to the Cucurbitaceae family and some varieties can be easily grown indoors. Those varieties with high-value markets are mostly exported to Japan and other Asian markets. The most popular varieties grown indoors are:

- netted melon (*C. melo* cv. *Reticulatus*)
- winter melon (honeydew melon; *C. melo* cv. *Inodorus*)
- cantaloupe melon (rock melon; *C. melo* cv. *Cantalupensis*)
- snake melon (*C. melo* cv. *Flexosus*)
- Oriental sweet melon (*C. melo* cv. *Makuwa*).

Mushrooms

The production of mushrooms is very much a niche industry in New Zealand, involving highly specialised growing conditions. It is also a growth industry, and in recent years there has been an upsurge in new varieties of mushroom (mostly of Asian origin) being introduced to New Zealand consumers. Aside from the fresh

market options there is a considerable market in processed and dried mushroom products. The most popular mushrooms now grown are:

- button or brown mushrooms (*Agaricus bisporus*)
- shiitake mushrooms (*Lentinus edodes*)
- oyster mushrooms (phoenix tails or pleurots; *Pleurotus pulmanarius*)
- enokitake mushrooms (enoki or golden needle mushrooms; *Flammulina velutipes*)
- truffles (*Tuber melanosporum*) — outdoor production only.

Commonly, mushrooms are produced in dark, climate-controlled rooms. Mushroom spores are collected and used to inoculate the growing medium that then becomes the spawn. The spawn is applied to compost and the cycle of production begins.

Sprouted beans and seeds

This classification of vegetable was previously categorised under the name 'bean sprouts'. There is a consistent demand for sprouted beans and seeds. The plants are harvested at such an immature stage of their life that they have a limited lifespan and require good post-harvest management. The key crops used to produce these products are:

- alfalfa
- lentils
- chickpeas
- adzuki beans
- kaiware (daikon radish seeds)
- sunflowers (shoots)
- mustard
- mung beans
- snowpeas
- blue peas
- cress
- broccoli
- soybeans.

Indoor production practices

In any indoor system the advantage for the producer is the ability to manipulate the environment to produce crops to a predetermined criteria and for a market. By managing the environmental factors, growers can work their system to specific harvest requirements including dates and volume. In any modern glasshouse or greenhouse system, growers now have access to computer programs that can be used to respond to production parameters such as climatic and water determinants. It is relatively easy to use computer systems to manage temperatures, airflow, nutrient additions and so on. For growers, the cost of investing in the establishment of an indoor production unit is set against the returns it is likely to generate.

The other mitigating factor for growers is that some crops respond better to indoor systems than others. Many culinary herbs can be managed easily in indoor systems; similarly, the classic salad crops of lettuce, tomatoes and cucumbers all have varieties that have been developed specifically for indoor systems.

For all indoor systems the seedlings are established as transplants in a separate unit prior to being put into the production unit. Seedling producers need to have specialised germination rooms, facilities for pricking out and growing on seedlings, etc. and, most importantly, they must practise strict hygiene to ensure high quality plants.

Within indoor systems there are some issues which cannot be ignored. The presence of some pest and disease pathogens that adapt to these systems particularly well can be an ongoing issue. The development of IPM programmes and stringent monitoring is important to maintain the plants at high levels of health. Note that some disease issues are physiological, resulting from nutritional or environmental factors (e.g. blossom-end rot of tomatoes and other fruits as a response to calcium levels).

Some of the more prevalent pests and diseases that might occur in an indoor system include:

- whitefly (*Trialeurodes vaporariorum*)
- blossom-end rot — tomatoes, capsicums
- aphids — many species
- botrytis (*Botrytis cinerea*)
- powdery mildew (*Sphaerotheca fuliginea*).

Outdoor production

Outdoor vegetables are predominant in New Zealand's horticulture industry. The three most important relative to their market values (onions, squash/kabocha and potatoes) are grown over extensive land areas in regions suited to their environmental needs and harvestability. The main outdoor vegetable crops produced in New Zealand are:

- onions
- squash/kabocha
- potatoes
- peas
- sweetcorn
- pumpkins
- watermelons
- kūmara (sweet potato).

There are many other crops grown outdoors. Some have a limited market focus (such as parsley and Brussels sprouts) and others are export-orientated (such as asparagus).

All vegetable crops in outdoor production systems require management programmes for weed, pest and disease issues. It is better for growers to be proactive in using protection measures rather than applying reactive management measures.

We have already introduced a range of vegetable crops that are grown in indoor systems; many can also be grown in outdoor systems during the preferred seasons, an option which can be much cheaper for producers. As with indoor crops, the establishment of the crop needs to be considered, whether it be by sowing in situ or through the use of transplants. There are specialist nurseries for transplant production (both bare-root and cell transplants). There is also a considerable amount of specialist machinery available for ground preparation, transplanting and crop maintenance through to harvest and post-harvest procedures.

Vegetables suited to outdoor production

Onions (*Allium cepa*) and other *Allium* crops

Onions are the highest export-earning vegetable crop, with exports mostly to European and Pacific countries. Related *Allium* crops include:

- onions (*Allium cepa*)
  - dry bulb (white) onions, pickling onions, silver-skinned onions, red-skinned onions and salad onions
- leeks (*Allium porrum*)
- spring onions (*Allium fistulosum*)
- garlic (*Allium sativum*)
- chives (*Allium schoenoprasum*)
- shallots (*Allium ascalonicum*).

All alliums belong to the Lilliaceae family and are grown from seed, generally sown in situ. The exception is garlic, which is grown from bulbils; the plant forms a set of bulbils known as a 'clove', which is harvested and then dried or cured. For onions and shallots it is important that the plants form a bulb, and this is a physiological stage as a response to certain environmental conditions, specifically day-length and temperature. Seed-bed establishment and weed control are very important factors in producing good allium yields. Other immature varieties such as spring onions and leeks are produced for local market in all regions of New Zealand.

Squash/kabocha (*Cucurbita maxima*)

Kabocha is a variety of winter squash and is the name applied to 'Japanese squash', a major export crop for New Zealand with over 8200 tonnes produced annually for markets in Japan and Mexico. It has a yellow-green flecked rind with yellow flesh and can be stored for about three months. The main production areas for kabocha squash are Hawke's Bay, Manawatū and the Rangitīkei and Whanganui regions.

Potatoes (*Solanum tuberosum*)

The modern European potatoes originated in the temperate regions of South America; these are now the most favoured vegetable in New Zealand and the fifth most-produced crop in the world. Modern varieties of potato are capable of producing marketable yields of up to 100 tonne/ha; however, most varieties produce somewhere between 40 and 90 tonne/ha. Potatoes are produced for fresh, export and processing markets with the key production areas in New Zealand being around Pukekohe in South Auckland, Ōpiki in the Manawatū, the Rangitīkei district, Canterbury, and South Canterbury and Otago for early season 'new' potatoes.

Potatoes are grown from 'seed' potato tubers, which are in fact a cutting of the parent plant rather than a 'seed'. An industry exists in New Zealand around 'seed' potato production, which is managed under the seed certification scheme run by the New Zealand Seed Potato Certification Authority (www.potatoesnz.co.nz/seed_potato_certification.html). Seed production is primarily based around the Canterbury–South Canterbury area, and the objectives of the scheme are to ensure that cultivars are as true to type as possible and meet pest and disease tolerances. Participation in the scheme is voluntary, as there is no compulsion for vendors, proprietors or agents of a seed tuber production venture to use the services of the authority.

There are many cultivars of potato grown for our markets, which are generally classified under three groups:

- **Early season** — e.g. Ilam Hardy, Swift, Jersey Benne, Cliff's Kidney; early-maturing in a period of up to 3 months (90 days) to harvest. Best eaten as 'new' potatoes and not stored.
- **Mid-season** — e.g. Desiree, Maris Anchor, Nadine; mid-season crops that take 4–5 months to reach maturity and can be harvested as either immature 'new' potatoes or full-term crops.
- **Late season** — e.g. Rua, Agria, Red Rascal; up to a maximum of 6 months to harvest. These varieties generally store well and are better suited to long-term storage.

Varieties are further differentiated depending on their end use, e.g. chipping or processing, and on their suitability to regional conditions.

Peas (*Pisum sativum*)

The pea is a green, pod-shaped vegetable, widely

grown as a cool-season crop. Generally peas are grown outdoors during the winter months in slightly acidic, well-drained soils. Most peas grown in New Zealand are used for processing rather than consumed as a fresh product, in large part because of their relatively short season. Processed pea exports exceed $84 million and are produced on over 8,000 ha of planted area throughout New Zealand. An interesting point to note is that processing peas are produced within a well-defined radius of the processing plant to ensure that the quality of the harvested crop is maximised from harvest through to freezing. As the harvest window for peas is very narrow, crops are planted sequentially to fit in with factory schedules and harvester capabilities.

Snow peas (*Pisum sativum* var. *macrocarponi*) are a 'flat' pea variety and are becoming increasingly popular, often used in salads and Asian cooking.

Sweetcorn (*Zea mays* var. *saccharata*)

Corn or maize is native to Central America but is now grown throughout the world. A sweet form of maize was developed, giving the name sweetcorn. Several varieties are available for commercial production, including some with white kernels and others with a mix of yellow and white kernels. Sweetcorn requires a long summer and warm temperatures to set properly and reach harvest standards. The main production areas are therefore located in areas where summers are long and dry and the land contours suit mechanised planting and harvest. Gisborne, Hawke's Bay and Canterbury are currently New Zealand's main sweetcorn-producing regions.

Commercial varieties differ in sweetness; recently the super-sweet varieties have become popular, with Honey 'n' Pearl being the most widely grown. Exports of processed and frozen sweetcorn from New Zealand total over $40 million annually. In New Zealand there are also some old, pre-commercial varieties of corn (not necessarily sweet) called kaanga by Māori and grown for processing and local consumption.

Pumpkins (*Cucurbita pepo*)

Pumpkins are grown in New Zealand for both the fresh market and processing. Some of the related 'squash' varieties are very similar to pumpkins and are produced and marketed alongside these crops. Varieties are grown to fit with local environmental conditions and storage requirements. Consumers tend to have preferences for specific varieties. There are many types of pumpkin, of which the most common are:

- butternut (*Cucurbita moschata*)
- Hubbard varieties (hard, warted and orange/red rind)
- Whangaparaoa Crown and Queensland Blue — both grey-green varieties with orange flesh and long-keeping qualities
- Jack O'Lantern — large fruit with smooth orange skin and orange flesh.

Butternuts are produced commercially as the basis for processed pumpkin products, especially soups. All are grown from direct-sown seed as the plants do not handle transplanting well. A good plant will yield 3–4 fruit each, although specific varieties will produce different yields. One of the key advantages of producing pumpkins is the ability to store them over several months if they are harvested at the right maturity and the stalk end is cured before storage.

Watermelons (*Citrullis lanatus*)

The watermelon is a native of tropical Africa and is widely grown around the world. Crops

are grown from seed and require warm, frost-free areas to achieve maturity. The Gisborne, Hawke's Bay and Bay of Plenty regions are well known for watermelon production.

It is important to manage crops for weeds, to minimise competition, and for pests and diseases, especially in a long-term rotation system.

It is not easy to determine the maturity of watermelon fruit; four methods are generally employed:

1. When the tendril that has developed from the same leaf axis as the fruit turns black.
2. When the groundspot on the fruit has turned a particular shade of yellow.
3. When the skin of the fruit gets a dull, rough appearance.
4. Tapping the melon to hear the resonance.

Seedless varieties of watermelon are now available.

Kūmara (*Ipomoea batatas*)

The kūmara or sweet potato is native to the tropical parts of South America; the sweet potato was domesticated there at least 5000 years ago. Sweet potatoes were grown before Western exploration in Polynesia, including New Zealand, where it is known as kūmara. As opposed to the common potato which forms harvestable stem tubers, kūmara grows on a creeping vine and produces swollen storage roots which are then harvested. The majority of commercial kūmara crops in New Zealand are grown in Northland in and around Dargaville and Ruawai, where the soil type and climatic conditions suit kūmara perfectly.

Note that because of the climatic limitations, kūmara does not flower in New Zealand. Crops are established annually, being produced from cuttings grown from tubers saved from the previous season. When harvested and cured properly, kūmara can be stored for many months and so are available all year round.

The most common kūmara variety grown in New Zealand is Owairaka Red — red skinned with creamy white flesh. Gold kūmara (also sold as Tokatoka Gold) has golden skin and flesh and a sweeter taste.

Other crops in New Zealand

Taewa or Māori potato (*Solanum tuberosum* ssp. *andigena*)

The taewa or Māori potato is known by a number of other names, and there are a number of different beliefs regarding their origin in New Zealand and the route they took to get here. Many people believe that there were cultivars of taewa in Aotearoa before European explorers such as Cook made contact.

Captain James Cook is credited with the earliest *recorded* introduction of potatoes to New Zealand. On his first voyage in November 1769 he visited Mercury Bay in the Coromandel region. Te Horeta Te Taniwha was a child at the time, but his recollections in old age included the following: 'Cook then gave two handfuls of potatoes to the old chief [Toiawa], a gift of profound importance to the Maoris. By tradition these potatoes were planted at Hunua where, after cultivation for 3 years, a feast was held and a general distribution made' (Begg and Begg 1969:36).

It is generally accepted that taewa were not brought as cargo during the migrations of Māori to Aotearoa, but how they arrived is an interesting point. Some believe that chance visits by trading vessels (unrecorded) that had earlier visited South America were responsible for the introduction of taewa. Lieutenant King, the governor of Norfolk Island, is known to have been a catalyst in the introduction of a range of exotic flora and fauna to the northern districts during a visit to New Zealand in 1793. Aside from presents of tools and implements, King is credited with the introduction of the European or 'white' potato, sometimes called rīwai.

Taewa or Māori potato ultimately replaced (or displaced) traditional crops such as kūmara and aruhe (fern root) as the primary carbohydrate and subsistence crop produced by Māori for their own use. Today taewa are produced using the same processes and technology as commercial potato crops, and have become a well recognised and distinct vegetable for many consumers.

Kamokamo — Cucurbitaceae family (believed to be *Cucurbita pepo*)

Kamokamo is a variant of the cucurbita, similar to the marrow *(Cucurbita pepo)*. It is a fast-growing summer-annual plant believed to have been introduced during the early years of settlement in the eighteenth and nineteenth centuries.

There are several distinct varieties of kamokamo, which have probably arisen from cross-pollination, seed selection and isolation over several decades. For many people kamokamo is a favourite vegetable, generally eaten as an immature fruit. It is sometimes marketed as kumikumi. Early crops are harvested from December onwards around the Bay of Plenty and later crops will fruit through to late summer in more-southern districts.

Pūhā (sometimes given as pūwhā) (*Sonchus* spp.)

Essentially not a *cultivated* crop for early Māori but one sourced from wild populations, pūhā (sow thistle) is experiencing a renaissance in the New Zealand diet and is now being cultivated purposefully for the market. It is generally cooked in the same way as spinach, added to soups or used raw in mixed-leaf salads. New Zealand is the only country in the world that utilises this plant as a commercial vegetable.

Pūhā is a wild vegetable that grows prolifically throughout New Zealand and off-shore islands. Even though these plants grow on all continents of the world, pūhā is considered to be an indigenous vegetable or food by most New Zealanders and is now under scrutiny as a potential commercial crop option. It is available at most farmers' markets and flea markets and some selected supermarkets. Trials on seedling and plant production are continuing.

A selection of kamokamo varieties.

Watercress (*Nasturtium orientalis*)

Watercress is an introduced plant, but one recognised as being important in the New Zealand diet and now being grown commercially by horticulturists throughout the country. Watercress is said to have originally been introduced by missionaries in the Bay of Islands in the early nineteenth century, and later (in 1841) to Canterbury by the sailing ship *Compte de Paris.* The commercial variety is generally produced from *Nasturtium officinale* seed sourced from countries such as Canada where watercress has been a commercial crop for many years. Non-commercial harvests of watercress are often sourced from rural waterways, where it establishes easily and grows for around 10 months of the year. In this situation, watercress can be harvested several times before the plant becomes exhausted. Currently watercress is only used as a fresh green vegetable.

Kōkihi or New Zealand spinach (*Tetragonia tetragonioides*)

Kōkihi is an indigenous plant found growing wild throughout the country, including the Kermadec Islands; mostly in coastal areas with sandy soils. Commonly called New Zealand spinach or perpetual spinach, it is not often seen in gardens or on dinner tables today yet was a popular vegetable with earlier generations of New Zealanders. The early explorers Cook and de Surville both utilised it as a substitute remedy for scurvy among their crews. From seeds collected on Cook's expeditions, kōkihi is now grown as a vegetable in many parts of the world. An interesting note regarding kōkihi is that it is believed to be the only true vegetable that any part of Australasia has provided to the world's cuisine, and yet it remains relatively unappreciated in New Zealand itself. This is another crop undergoing a resurgence in interest, and it is possible to purchase the seeds for home gardens and occasionally harvested plants for cooking.

Organic vegetable production

Conventional vegetable production depends to varying degrees on the use of chemical fertilisers and pesticides. Organic vegetable production does not involve these chemicals or genetically engineered plants. Conventional production systems are generally considered to be more damaging to the environment than organic systems. The former have been associated with soil compaction, reduced soil organic matter and biological activity, water pollution with pesticides and chemical fertilisers, and other negative impacts on the environment.

The fundamental principles underlying organic production include natural plant nutrition, natural pest control and biodiversity. The system is sustained by using green manures, animal manures, mulches and crop rotations (Welbaum 2015). Organic production systems aim to improve soil characteristics and eliminate the use of synthetic chemicals.

Organic production allows small-scale producers to differentiate their product from that of conventional growers and thus claim a price differential. However, larger-scale producers have adopted organic principles and it has become increasingly accepted as a commercial production system.

The arguments about whether organic

farming systems are sustainable is ongoing. Generally organic systems are less productive, more labour-intensive and more complicated than conventional systems and operate without any of the immediate chemical supports used by conventional producers. Moreover, debate also continues about whether organically produced crops are healthier for consumption than conventionally produced crops.

In New Zealand the standards that affirm that a grower is organic are controlled by several organisations, including BioGro, Biodynamics and the government.

References

Begg, A.C. & Begg, N.C. 1969. *James Cook and New Zealand*. Wellington: A. R. Shearer.

Fresh Facts. 2019. *Fresh Facts 2019*. Accessed 2 September 2020. www.freshfacts.co.nz/files/freshfacts-2019.pdf.

Welbaum, G.E. 2015. *Vegetable production and practices*. Wallingford, Oxon., UK: CAB International.

Research

Vegetable production is an important part of the overall food production for humans. It is constantly adapting to allow for changing needs, to meet environmental changes and to cope with new, recently introduced pests and diseases. Moreover, there is a constant need to make vegetable production more efficient with regard to water, fertiliser and chemical use.

One standard way of controlling inputs and outputs is hydroponic systems. More recently, indoor vertical production systems that allow concentrated production indoors of salad crops have become available.

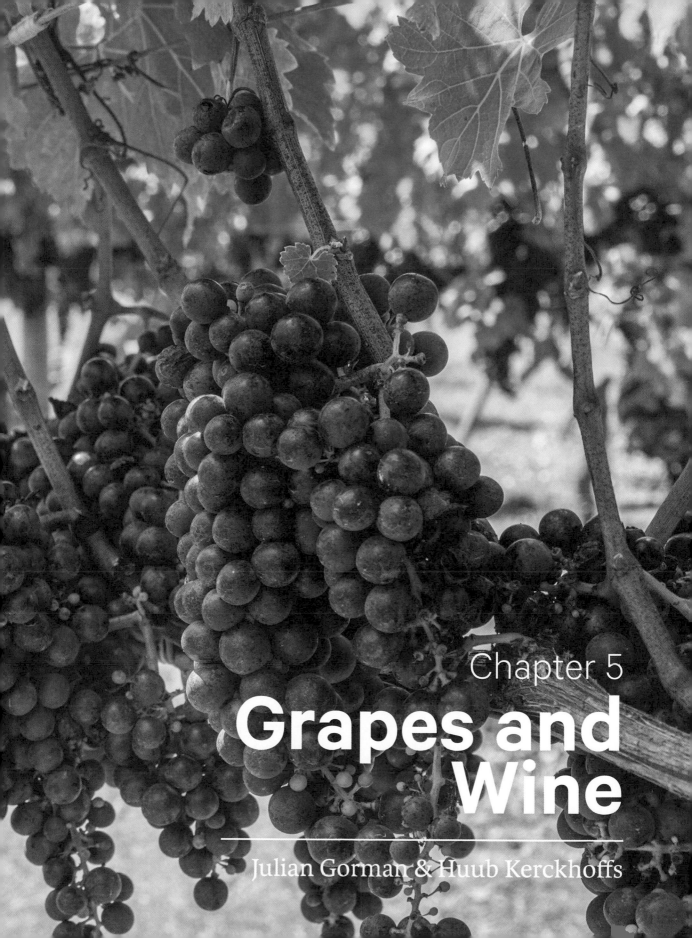

Chapter 5

Grapes and Wine

Julian Gorman & Huub Kerckhoffs

Chapter 5

Grapes and Wine

Julian Gorman

School of People, Environment and Planning, Massey University

Huub Kerckhoffs

Horticultural Sector Policy, Ministry for Primary Industries

Introduction

The development of the New Zealand grape and wine industry is a remarkable story of success. According to New Zealand Wine (2020), the history of the industry extends back to the 1800s. But it was the introduction of the Marlborough region's astonishing sauvignon blanc in the 1980s that saw New Zealand wine explode on to the international scene. While Marlborough retains its status as one of the world's foremost wine-producing regions, the quality of wines from elsewhere in the country has also garnered international acclaim. By world standards, New Zealand's production capacity is tiny, the country's total volume of 302 million litres accounting for less than 1 per cent of total volume globally.

The average price tag of a New Zealand wine is a reflection of its desirability, and few question its ability to deliver excellent value for money. In fact, it is ongoing commitment to quality over quantity that has won New Zealand its reputation as a premium producer. This success is thought to have arisen from the combined effects of a temperate maritime climate, the passion of the wine producers and the very distinctive nature of New Zealand's wine styles (Care 2016). New Zealand wines in general achieve premium prices in export markets and are competitive with premium wines from other temperate-climate countries. Of the wines from New Zealand, the highest export demand is for sauvignon blanc, with other wines, such as merlot, starting to gain ground and provide additional opportunities for

New Zealand's wineries (Ministry for Primary Industries 2019).

This chapter explores the history and origins of the grape and wine industry, along with the reasons for its success, and touches on aspects of viticulture and winemaking, and future challenges and opportunities for wine production in New Zealand.

The grape industry in New Zealand

At a global level, New Zealand is a small wine producer with an annual production of less than 1 per cent of the world's wine. However, it has very rapidly gained a strong demand in the export market for a range of high-quality and different styles of wines. In 2019, New Zealand wine was exported to over 100 countries, earning more than $1.8 billion. Over 80 per cent of this was exported to four countries: the United States,

$550 million; the United Kingdom, $441 million; Australia, $368 million; and Canada, $129 million (Figure 5.1). In the same year, New Zealand imported wine to the value of NZ$225 million from 40 countries, with Australia and France accounting for 75 per cent of this value. Also important to New Zealand's winegrowers are the 750,000 international wine tourists who in 2018/19 spent NZ$3.2 billion visiting wineries as part of wine-tasting tours, dining and accommodation (New Zealand Winegrowers 2019a).

The Marlborough region is the biggest grape-growing region, with 69 per cent of the total vineyard area; Hawke's Bay, Central Otago and Gisborne are also significant but much smaller, with areas of 12, 5 and 3 per cent respectively. According to New Zealand Wine (2020), New Zealand has 38,073 hectares (ha) in grapes, with 30,092 ha growing white wine grapes and 7876 ha growing red wine grapes, across 2031 vineyards, with an average size of 18.5 ha. In 1983, Müller-Thurgau was the most planted white grape variety and cabernet sauvignon the most

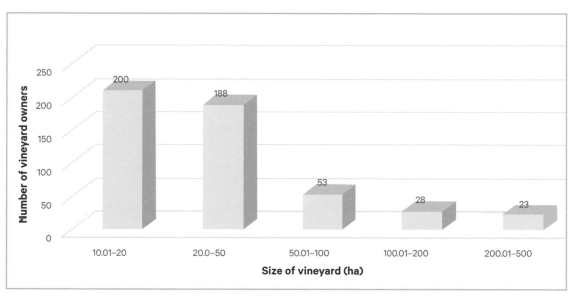

Figure 5.1 Number of vineyard owners vs vineyard size (ha), 2018. (Source: New Zealand Winegrowers 2019a)

Table 5.1 Top wine types produced in New Zealand today.

Red varieties		White varieties	
Pinot noir	72%	Sauvignon blanc	77%
Merlot	15%	Chardonnay	11%
Syrah	6%	Pinot gris	8%
Cabernet sauvignon	3%	Riesling	2%
Malbec	2%	Gewürztraminer	1%
Cabernet franc	1%	Other	1%
Other	2%		
Total:	**7876 ha**	**Total:**	**30,092 ha**

Source: New Zealand Wine 2020

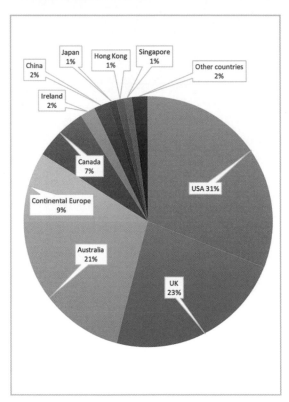

Figure 5.2 New Zealand wine exports by country, 2019. (Source: Fresh Facts 2018)

planted red variety. Since 1990 there has been a definite change in grape varieties planted in New Zealand. Sauvignon blanc is now the most widely planted white variety, with chardonnay following in second place (Table 5.1). Other white varieties such as pinot gris, riesling and gewürztraminer have dramatically increased in production level and lesser-known varieties such as Viognier are now being grown in small quantities. A surge in the production of pinot noir has seen this variety become the most widely planted red variety in New Zealand. Syrah (shiraz) is another red variety that has grown considerably, with merlot plantings remaining strong, though no longer increasing (New Zealand Wine 2020).

New Zealand's wine industry had a massive growth spurt in the 2000s, with the number of wineries increasing from 358 in 2000 to 672 in 2010 and the total grape-producing area going from 10,197 ha to 33,425 ha over the same period (Care 2016). The export value of the wine industry grew substantially between 2000 ($167 million) and 2010 ($1041 million), when it became one of New Zealand's top 10 export earners. However, there was a downturn during the global economic crisis in 2008, when the average export value of grapes dropped significantly. This had significant impacts on the land value of vineyards. An international oversupply of wine, brought about by falling global sales, slowed the development of new vineyards.

The diversification of grape varieties has decreased over time, with more of a concentration of the sauvignon blanc variety. This increased from 39 per cent of the total producing area in 2006 to 58 per cent in 2014. Vineyard growth levelled out between 2012 and 2014, but is thought to be increasing again following a trend that is seeing larger corporate entities becoming more dominant. This is a different driver of vineyard-area growth than in the early 2000s, which were predominantly small investors developing 10- to 20-ha blocks (Care 2016). Figure 5.2 shows the breakdown of vineyard owners by size of vineyard in 2018, indicating that 93 per cent of vineyard owners have a vineyard less than 100 ha in size.

History of winemaking in New Zealand

For much of the twentieth century, grape growing in New Zealand was confined to the warm top half of the North Island; and it was not until the 1970s that vineyards started being established south of Hawke's Bay. Until the 1960s, the government advised against establishing vineyards in the South Island as it was deemed too cold and unsuitable for growing wine grapes. However, in the 1970s and 1980s, growers trialled new varieties with great success, especially in the Marlborough and Otago districts (see Figure 5.3). Winemaking in New Zealand began in the nineteenth century,

mainly by immigrant winemaking families. One of the earliest vineyards was at Waitangi, in James Busby's garden. Another well-known early vineyard was The Mission in Hawke's Bay, run by Society of Mary priests and brothers. Many Dalmatian immigrants developed vineyards around Auckland (e.g. the Babich and Fistonich families) after they left the gum fields of Northland in the nineteenth century.

The species grown was *Vitis vinifera*, a European species. However, traditional European grapes were not suitable in New Zealand as the climate was very different, with high rain and humidity. These climatic conditions caused greater losses of grapes on sites around Auckland, and the remaining crop did not produce good-quality wine (Jackson and Schuster 1994).

In 1902, the New Zealand government commissioned the services of an Italian viticulturist, Romeo Bragato, to advise the wine industry. Romeo had trained at the then highly regarded School of Viticulture and Oenology in Conegliano, Italy, between 1878 and 1883. He had also been employed by the Victorian Government of Australia as Viticulturist Expert between 1888 and 1901. Bragato moved to New Zealand in 1902, where he was employed as the head of the Viticultural section of the New Zealand Department of Agriculture until 1909. Bragato was a visionary and was responsible for some of the most important developments in New Zealand's grape growing. He identified the regions that were suitable for wine growing, identified appropriate vines and varieties, diagnosed phylloxera disease in Auckland vineyards and recommended the importation of cuttings of phylloxera-resistant American rootstock. Bragato showed growers how to graft European classical varieties, designed vineyard layouts, developed pruning methods, and much more. In recognition of his contribution to the industry, a Regional Research Institute was named the Bragato Research Institute in his honour (Bragato Research Institute 2019).

In the 1950s there was a dramatic increase in the demand for sparkling and fortified wines (ports and sherry), which resulted in vineyards being planted in the Auckland and Hawke's Bay areas. Most of the wine produced was of average quality, but there were several fine table wines which showed the potential. During the 1950s and 1960s, new technology, such as temperature-controlled fermentation, was introduced. Alex Corban from the University of Auckland pioneered winemaking techniques and introduced them through Corbans Wines, using modern stainless-steel fermentation tanks.

The 1960s and 1970s laid the foundations for the modern grape and wine industry in New Zealand. At this time a so-called hybrid grape was grown because it had high yields and was hardy and disease-resistant. However, the resulting wines were of low quality and were only acceptable as fortified wines. The 1970s saw a rapid expansion in the New Zealand wine industry with several large companies gaining in size by merging, resulting in an increase in the volume of wine of better quality and variety.

In the 1980s Peter Hubscher of Montana Wines produced a world-class sauvignon blanc which won awards around the globe, and today this variety is the standard-bearer for local production. It is widely grown in Marlborough. Advances such as these were guided by the New Zealand Winemaking Industry Committee, which promoted knowledge-intensive production systems (Barker, Lewis, and Moran 2001).

Subsequently, varieties of good-quality grapes that are better suited to New Zealand have been selected for wine production, and in the past 20 years the sector has boomed, with strong growth in exports and wine becoming a major sector in New Zealand's horticultural industry.

Grape cultivation in New Zealand

The European grape, *Vitis vinifera*, the main species grown in New Zealand as a wine grape, probably originated in the Caucasus Mountains between the Black and Caspian Seas. Internationally this species is used for much of wine production and table grapes. The other species, *V. labrusca*, which is of American origin, is only grown as a table grape in New Zealand (e.g. Albany Surprise). Of 8000 known grape varieties, only 50 are grown in New Zealand.

In the early years of the modern grape/wine industry in New Zealand, the 1960s and 1970s, hybrid grapes of *V. vinifera* and *V. labrusca* were grown. Although these were high-yielding, disease-resistant and hardy, they made poor-quality wines that were only acceptable for fortified wines (ports and sherries) (Hira and Benson-Rea 2013). The expansion in grape growing in the 1960s and 1970s resulted in wine of low quality for which the market was limited.

In the mid-1980s, the industry convinced the New Zealand government to compensate growers for pulling out old vineyards of poor grape quality, enabling growers to plant preferred varieties. This facilitated the emergence of premium wine styles that were focused on both local and export markets. The growth indicators for the New Zealand wine industry subsequently showed massive growth between 1990 and 2013,

Table 5.2 New Zealand wine industry key growth indicators (1990–2018).

Indicator	1990	2013	2018	% change (1990–2018)	% change (2013–2018)
Number of wineries	131	698	697	432% ↑	0.1% ↓
Producing area (hectares)	4880	35,182	37, 969	678% ↑	7.9% ↑
Grape production (tonnes)	70	345	419	499 ↑	21.4% ↑
Average yield (tonnes per hectare)	14.4	9.8	11.1	22.9% ↓	13.3% ↑
Wine exports (million litres)	4	170	255	6275% ↑	50% ↑
Wine exports (NZ$ million)	18.4	1210	1705	9166% ↑	40.9% ↑

Sources: Hira and Benson-Rea 2013; Plant and Food Research 2018; New Zealand Wine 2020

with much more modest growth between 2013 and 2018 (Table 5.2). Major foreign investment brought in new techniques, equipment and marketing knowledge, along with some necessary restructuring of the industry (Barker, Lewis, and Moran 2001). In 2004, for the first time, the New Zealand wine industry exported more wine than was consumed domestically (Brodie, Benson-Rea, and Lewis 2008).

New Zealand has followed the lead of other competitors in the global wine industry (such as Australia) by building a New Zealand wine brand. The largest volume of wine produced is sauvignon blanc, followed by pinot noir, chardonnay, pinot gris, merlot and riesling (Plant and Food Research 2018). Marlborough is the biggest wine-producing region, followed by Hawke's Bay, Otago, Gisborne, Canterbury/Waipara, Nelson and others (Figure 5.3, Table 5.3).

New Zealand wines

New Zealand wines are unique and have put the country on the world map for wine.

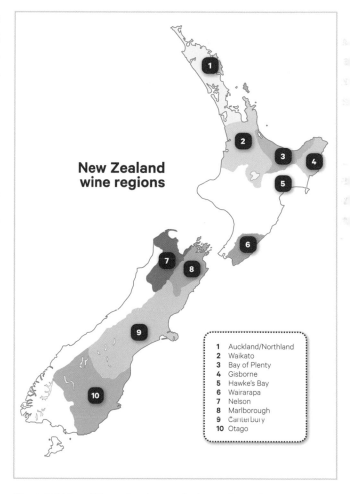

Figure 5.3 Map of the main wine-producing regions of New Zealand. (Source: New Zealand Wine 2020)

Table 5.3: Features of New Zealand's major wine-growing areas.

Region	Features	Rainfall (mm)	Largest variety
Auckland, 314 ha	Produce red wines that grow on shallow clay soils over sandy clay or silty clay subsoils. Auckland has notable cabernet sauvignon and chardonnay.	1500–1600	Chardonnay 79 ha
Northland, 78 ha			Chardonnay 20.7 ha
Waikato/ Bay of Plenty, 15 ha	Sunny but wet climate which is not well suited to grape growing. Heavy loams over clay subsoils.	1100–1200	Pinot noir 5.7 ha
Gisborne, 1190 ha	Fourth ranked grape-growing region in New Zealand. Excellent chardonnay, chenin blanc, gewürztraminer and riesling wines. Fertile alluvial loams over sandy or volcanic subsoils.	1000–1050	Chardonnay 594.2 ha
Hawke's Bay, 4771 ha	Mild weather and diverse soil composition. Vineyards grow chardonnay, cabernet sauvignon, merlot, Viognier, cabernet franc and syrah (shiraz) grapes. Clay loams of medium to high fertility over gravely or volcanic subsoils.	750–800	Sauvignon blanc 1104.6 ha
Nelson, 1154 ha	Soil composition is suitable for growing various types of wine; region includes the sunniest city in New Zealand. Clay loams over hard clay subsoils.	1000–1250	Sauvignon blanc 617 ha
Marlborough, 26,850 ha	Biggest wine-growing district in New Zealand. World's best sauvignon blanc, first-rate chardonnay, riesling and pinot noir. Silty alluvial loams over gravelly subsoils. In parts compacted silt or clay pans of various thickness and depth.	650–750	Sauvignon blanc 21,415 ha
Canterbury, 157 ha	Dry summer and cool winter. Stony and alluvial soils. Alluvial silt loams over gravel subsoils in central parts. Chalky loam soils often rich in limestone in the northern part.	600–750	Pinot noir 60 ha
Waipara Valley, 1226 ha	Long sunshine hours, low rainfall, cool nights and free-draining soils. Produces best grape varieties in New Zealand.	600–750	Pinot noir 370 ha
Central Otago, 1884 ha	Southernmost wine-producing region in the world. Hot and dry summers and snowy winters; produces outstanding grapes. Silt loams with mica and schists.	400–450	Pinot noir 1501.7 ha
TOTAL PRODUCING AREA 37,639 ha			

Sources: Care 2017; New Zealand Wine 2020

The unique characteristics of the wine of New Zealand's regions can now be formally recognised through registration of a region's name under New Zealand's Geographical Indications (Wine and Spirits) Registration Act 2006 (Plant and Food Research 2018). One hundred per cent of the wine must be from New Zealand and at least 85 per cent from the region, variety and vintage that appears on the label of the wine.

Factors in New Zealand's increase in wine production

- The introduction of early-ripening grape varieties being well suited to the main growing districts (Gisborne, Hawke's Bay and Marlborough);
- Modernisation of winemaking techniques and production methods;
- Improved disease control;
- Planting of improved material through regional Vine Improvement Groups;
- Expansion of grape growing in the cooler districts of Hawke's Bay, Wairarapa and the South Island;
- Legislation controlling wine quality;
- Increased research and development investment in grape growing and winemaking;
- A strong focus on export and diversified markets. (Jackson and Schuster 1994)

A major replanting of premium grape varieties in the late 1980s marked a turning point in the New Zealand wine industry as it focused on the white wines (sauvignon blanc, chardonnay and riesling) for which there were recognised international standards. In addition, cabernet sauvignon, merlot and pinot noir were produced successfully in the southern regions (Jackson and Schuster 1994).

The dominant wine companies involved in the development of the New Zealand wine industry included Montana, Corbans, Nobilo and Selaks. These companies have their origins in mostly Croatian immigrant families, but most are now owned by large foreign multinationals. Only one large winery, Villa Maria, is still New Zealand-owned (Hayward and Lewis 2008), though another winemaker, Delegat Limited, is a New Zealand company controlled by its founding family of the same name. While the number of wine companies increased from 358 to 698 between 2000 and 2011 (National Bank of New Zealand 2012), the six largest companies in New Zealand accounted for approximately 55 per cent of total wine production and 19 per cent of grape production (Scandurra 2011).

Overseas owners now control about 40 per cent of New Zealand's wine production (National Bank of New Zealand 2012). Foreign investment has enabled New Zealand export growth in three new areas for commercial wine cultivation: Gisborne, Marlborough and Hawke's Bay (see Figure 5.3). By 2001, 82 per cent of New Zealand's vineyards were based in these three regions.

Structure and governance of the New Zealand grape and wine industry

The Wine Institute of New Zealand (WINZ) was formed in 1975 and when levy powers were granted in 1976 there was a fusion of state and industry interests to grow the industry. This led to the formation of a collective industry association which resulted in a more strategic development approach to growing this sector. The 1981–86 *Wine Industry Development Plan* laid the foundations for development of an export market (Hira and Benson-Rea 2013).

In the early 2000s, significant overlaps in both mandate and operations were identified between the two national organisations, the WINZ and the New Zealand Grape Growers Council (NZGGC). These organisations were merged in 2002 to become a single organisation, New Zealand Winegrowers, which was eventually incorporated in 2016. In its 2017 annual report, New Zealand Winegrowers set a target of $2 billion of export revenue by 2020. It believes

that this can be achieved due to New Zealand's international reputation for high-quality wines, a higher per-unit price, and the continuing increase in vineyard planting area. The area planted has grown from 7410 ha in 1997 to 37,129 ha in 2017, and currently New Zealand's wine exports are worth $1.8 billion (Plant and Food Research 2020).

Sauvignon blanc dominates New Zealand's wine sales and continues to increase in sales volume. Pinot noir, pinot gris and rosé, while sold in much smaller volumes, are also showing excellent growth rates (Table 5.4). Other varieties, such as chardonnay, riesling, cabernet sau-

Table 5.4 New Zealand wine industry — key growth indicators (1990–2018).

Indicator	1990	2018	% change
Number of vineyards	131	2019	1541% ↑
Producing area (hectares)	4,880	38, 073	680% ↑
Average yield (tonnes per hectare)	14.4	6.72	53.3 % ↓
Wine exports (million litres)	4	256	6300 % ↑
Wine exports ($million)	18.4	1,693	9101 % ↑
Indicator	**1990**	**2013**	**2018**
Number of wineries	131	698	697
Producing area (hectares)	4,880	35,182	37, 969
Grape production (tonnes)	70	345	419
Average yield (tonnes per hectare)	14.4	9.8	11.1
Wine exports (million litres)	4	170	255
Wine exports (NZ$ million)	18.4	1,210	1,705
Top export variety	**2009**	**2013**	**2018**
Sauvignon blanc (ML)	91.527	144.541	220.065
Pinot noir (ML)	6.183	10.17	13.171
Pinot gris (ML)	2.036	3.612	7.74
Chardonnay (ML)	4.789	4.914	4.766
Rose (ML)	0.706	0.49	3.656
Merlot (ML)	1.931	2.059	2.06
Sparkling (ML)	1.976	1.451	1.167

ML=millions of litres. Source: Hira and Benson-Rea 2013; FreshFact 2018; NZW 2019

vignon, merlot and pinot noir, have expanded, and in more-recent years pinot noir has made its mark as a New Zealand wine that is sought after in overseas markets and is commanding market premiums.

New Zealand Winegrowers is now the national industry body that represents the country's viticulture and winemaking sectors. It conducts research, promotion, marketing and advocacy in the interests of New Zealand grape growers and winemakers. Winemakers and grape growers are automatically entitled to membership of New Zealand Winegrowers through the payment of levies on grape or wine sales that are required by law in the Commodity Levies Act 1991 and the Wine Act 2003. New Zealand is the only country in the world with a single national wine industry body that represents and advocates for its entire grape and wine industry.

Horticultural practice

Climate and soils are important factors for growing grapes for wine, and the success of New Zealand's wine production can largely be attributed to the country's range of climates and soil compositions, including several alluvial-based soils that are ideal for growing a variety of grapes.

Climate

New Zealand's climate ranges from subtropical conditions at the top of the North Island down to a cold-temperate climate in the most southerly grape-growing regions of Central Otago (see Figure 5.3), spanning a distance of 1600 km. The overall New Zealand climate is characterised as maritime, often providing long sunshine hours alternated with nights with cool sea breezes.

The generally moderate temperatures allow for a long ripening period, which allows flavour development of the grapes while retaining a fresh acidity. This combination creates the intensity and vibrancy that New Zealand wines are renowned for. Many of New Zealand's vineyards are on the east side in the rain shadow of the ranges that run along the backbone of the country.

Soils

Grape vines for wine production are generally grown on low-fertile and free-draining soils. New Zealand soils have a strong influence on the style and character of its wine. Soils range from heavy, water-retaining clay loams in Northland to the dry, stony silts of the Wairau Valley and fertile flood plains (see Table 5.3). Grape vines prefer soils with high levels of organic matter that can store nutrients and water, neutralise toxins and stabilise soils, and decomposing material that can bind soils together. Soil management is necessary to enhance soil properties by increasing organic matter, reducing erosion, and improving soil structure and fertility. Vineyards need to develop fertiliser and nutrient management plans so that fertiliser application is in response to plant requirements to promote maximum uptake by the vines and minimise leaching and run-off.

Water

Water is very important to New Zealand's wine industry, being used for the irrigation of vines, frost protection and in wine production operations. Water use requires careful budgeting to ensure that there is enough irrigation for good fruit quality without overwatering. Viticulturists apply for water use through resource consent applications granted by local governments under the Resource Management Act 1991 (RMA). Their water usage, water drainage and wastewater disposal must be compliant with local or district area plans. In recent times viticulturists have had to improve the way they manage their water use, as access to water is becoming an important issue as more vineyards have become concentrated in certain areas.

Some frost-managing techniques use water, but these are set up to use the minimum amount possible. As frosts also tend to occur at times when there is high water supply, this practice is thought to have minimal impacts. Irrigation of vineyards is carefully monitored, and only applied according to plant needs and soil moisture levels through drip irrigation systems. In addition to saving water, this practice avoids the accumulation of excess soil water, which can lead to an increase in disease and a reduction in fruit quality as well as leaching and surface run-off that can degrade the quality of groundwater.

There are many parameters used to monitor water use in vineyards, including rainfall measurement, evapotranspiration and soil moisture measurements. Computer modelling and visual assessment is then used to make (daily) recommendations on water usage.

Other

There are some other important considerations, such as air quality, that need to comply with the RMA and the local regional plan or district/ unitary council plans. Issues that might require consents under such plans relate to spraying, vineyard heaters (for frost), noise, odour, buildings and conservation. There may be noise restrictions under Civil Aviation Authority rules if helicopters are used for preventing frost dam-

age. The Sustainable Winegrowing New Zealand Program led by New Zealand Winegrowers has standards and operational procedures in place that vineyard operations have to meet; these include site management plans, spray plans, spray diaries and monitoring records. Contractors administering spray need to have Growsafe training, which includes training on how to calibrate sprayers and safely apply spray.

The RMA is the principal item of legislation that affects the application of fertiliser. The Code of Practice for Nutrient Management 2007 provides a framework for the overall management of nutrients and a guide for best nutrient management practice and is supported by the Fertiliser Association, which encourages responsible and scientific-based nutrient management (Fertiliser Association 2020).

Establishing a vineyard

Selecting and establishing a vineyard site can cost between $25,000 and $50,000 per hectare, and it will take a number of years before the vineyard is established and providing an income. It is very important to choose and set up the site according to important criteria, such as the right rootstock and varieties to grow, and to set up the trellises in a way that will provide the best-quality grapes. Aspects to consider include:

- Site selection — soils, climate and accessibility are all important criteria in choosing a site for a vineyard.
- Stock selection — there are many varieties of grapes available, with different varieties being suited to different sites. There are also certified virus-free vines that are grafted onto phylloxera-resistant rootstocks.
- Planting and training vines — in New Zealand a wire trellis system is used to support vines

above the ground; the vines can be trained through pruning to maximise the amount of sunlight on their leaves to increase growth and reduce pests and diseases. Posts for trellises have in the past been treated pine; there is now a shift to use eucalypt hardwood, plastic or steel posts to avoid chemicals from tanalised wooden posts leaching into the water table.

- Young vines require water until they establish themselves, so a drip irrigation system is required.
- Frost can severely affect grapes, so there must be investment in fans and other methods to increase air flow to reduce frost damage.

In addition to the high establishment costs of vineyards, there are also high labour requirements as the vines need to be pruned and trained, the leaves plucked, and the grapes harvested.

Future issues, threats and opportunities

Future markets and growth predictions

The past 10 years has seen a rapid growth in the New Zealand wine industry, with a quadrupling of production and exports growing from 74.7 per cent of production to 90.6 per cent of production (NZ Wine Growers 2019a). The medium-term outlook for growth is expected to be more modest, as the industry has matured and there are fewer areas available for rapid expansion (Ministry for Primary Industries 2019). The total vineyard area is not expected to grow much (from 38,700 ha in 2019 to 40,000 ha by 2023).

Instead, a phase of consolidation is expected. Between 2010 and 2018 the average size of a vineyard increased from 29.4 ha to 54.3 ha and the number of grape growers dropped from 1100 to under 700 (Plant and Food Research 2018).

The Sustainable Winegrowers New Zealand programme is a voluntary, industry-wide initiative that was introduced in 1995 and developed to encourage an environmental 'best practice' model for vineyards and wineries in New Zealand. Sustainability has quickly become an important issue in New Zealand vineyards, with many turning to organic practices. Energy conservation and waste reduction initiatives are now key considerations, as are reducing water use and fertiliser leaching.

In 2008 Yealands Estate launched their vision of being the world's most sustainable winery. They have instituted many cutting-edge technologies and work practices. For example, they burn 10 per cent of their vine prunings to produce renewable energy, with the remainder getting mulched back into the soil. They have imported 'mini' sheep from Australia to eat the grass between rows — thus reducing mowing and herbicide spraying.

Many advances in winemaking technology have increased both efficiency and quality within the industry. Temperature control is a key part of the fermentation process and determines the quality of the wine. A New Zealand company that makes temperature and fermentation equipment has produced VinWizard, a central remote-control system, that can manage temperature remotely and save not only time but also from 30 to 50 per cent of energy costs.

The United States is New Zealand's biggest export market, and there are threats from increased tariffs for wine exported to the US.

Table 5.5 Annual cycle of the grape vine.

Late autumn (May)

Leaves turn yellow or red and fall from the vine.

Winter (June–late August)

Bare vines are pruned – most of the previous year's growth is cut away, leaving a few short shoots with buds.

Spring (September–November)

Buds swell and burst into leafy growth – clusters of flowers appear on new shoots and develop into bunches of grapes.

Summer (December–February)

Grapes grow and ripen.

Early autumn (March–May)

Grapes are harvested.

However, New Zealand has a healthy diversity of export markets, and New Zealand Winegrowers is constantly predicting future export demands along with a variety of other economic criteria in the event of additional tariffs being imposed.

Water quality

Water quality is a major ecological concern in New Zealand, and a key issue is the management of nitrates that run off and leach into groundwater tables from horticultural land use.

Labour issues

The grape industry is labour-intensive and suffers from a shortage of workers for pruning and picking. Seasonal peaks around harvesting and pruning time can require thousands of additional workers nationally. Growers are finding it difficult to source good-quality labour. Many vineyards are moving towards automated picking machines, but there are many jobs left that require human skills. Skilled labour programs such as the Recognised Seasonal Employer Scheme (started in 2007) are becoming more common and involve bringing in vineyard workers from the Pacific Islands.

Biosecurity

It is critical that the grape-growing sector be prepared to respond to anything that threatens the sector's long-term sustainability, and mitigating biosecurity risk is fundamental to the New Zealand wine industry. While quite geographically isolated, New Zealand is increasingly connected in a global world. Pests and diseases can be spread very rapidly, and a range of biosecurity threats exist from imported goods, passenger arrivals, climate change, increased volumes of exotic products, monocultures, and from goods accessed through internet shopping. New Zealand Winegrowers' biosecurity strategy sets out the biosecurity targets to be achieved each calendar year (New Zealand Winegrowers 2020).

Brown marmorated stink bug (BMSB)

This pest is not established in New Zealand but has spread from Asia to the United States and Europe and could be a significant economic pest as it feeds heavily on a wide variety of plant species. In July 2017 New Zealand Winegrowers signed the BMSB Operational Agreement, which sets up a partnership with other industry organisations and the Ministry for Primary Industries (MPI) to help improve the country's readiness for this pest.

Harlequin ladybird

This pest continues to spread through New Zealand's wine regions and is a threat to wine quality at harvest time. It is well established in Gisborne and Hawke's Bay and has been detected in Marlborough and Nelson. A management plan is required to help deal with this pest (New Zealand Wine 2020).

Phylloxera

The root-ruining phylloxera disease has been a major issue for grape growers in New Zealand and is largely responsible for a shift to grafted American rootstock species. Phylloxera was first detected in Auckland vineyards in the 1890s; growers were advised to destroy all crops harbouring the pest and replace the vines with grafted European grape varieties on phylloxera-resistant American rootstock — this practice had previously saved vineyards in Europe (Care 2016). A variety of strategies were followed by New Zealand growers over the years, but in the past century phylloxera continued to spread throughout New Zealand.

Most of the country's vineyard rootstock is now derived from three American species: *Vitis riparia*, *V. berlandieri* and *V. rupestris*, or their hybrid combinations (Care 2016).

Mildew

Powdery mildew can be a problem; in the 2014–2015 season there was an outbreak that affected a number of regions.

Frost

Frost is a problem for grape growers. Early autumn frosts can damage the grape crop just as it approaches harvest time, and late spring frosts can wipe out young shoots and flowers. Growers use a number of strategies to reduce damage from frost, with the most common being large fans or even low-flying helicopters to produce movement in the air and stop the frost forming.

References

Barker, J., Lewis, N., and Moran, W. 2001. 'Reregulation and the development of the New Zealand wine industry.' *Journal of Wine Research* 12 (3): 199–222.

Bragato Research Institute. 2019. 'About us.' Accessed 1 September 2020. https://bri.co.nz/about-us/.

Brodie, R.J., Benson-Rea, M., and Lewis, N. 2008. 'Generic branding of New Zealand wine: From global allocator to global marketing'. Paper presented at the Fourth Annual Conference of the Academy of Wine Business Research, Siena, Italy, 17–19 July 2008.

Care, D. 2016. 'An overview of New Zealand's viticultural and aquaculture industries.' Hamilton: Waikato Institute of Technology. Accessed 1 September 2020. http://researcharchive.wintec.ac.nz/4448/1/Overview%20of%20NZ%20Viticulture%20%26%20Aquaculture.pdf.

Fertiliser Association. 2020. Code of Practice. Accessed 17 September 2020. www.fertiliser.org.nz/Site/code-of-practice/.

Hayward, D., and Lewis, N. 2008. 'Regional dynamics in the globalising wine industry: The case of Marlborough, New Zealand.' *Geographical Journal* 174 (2): 124–37.

Hira, A., and Benson-Rea, M. 2013. 'New Zealand wine: A model for other small industries?' *Prometheus* 31 (4): 387–98.

Jackson, D., and Schuster, D. 1994. *The production of grapes and wine in cool climates*. Christchurch: Gypsum Press.

Ministry for Primary Industries. 2019. 'Situation and outlook for primary industries data', March 2019. www.mpi.govt.nz/news-and-resources /open-data-and-forecasting/situation-and-outlook-for-primary-industries-data/.

National Bank of New Zealand. 2012. *NBNZ Agri-Focus*, March, Report. www.anz.co.nz/resources/3/0/309e22004ab1bf7ab62afe415e15c706/ANZ-AgriFocus-20120330.pdf.

New Zealand Wine. 2020. *New Zealand Wine: A comprehensive guide to the regions and varieties*. Accessed 1 September 2020. https://253qv1sx4ey389p9wtpp9sj0-wpengine.netdna-ssl.com/wp-content/uploads/2020/06/NewZealand_WineTextbook_WebVersion.pdf.

New Zealand Winegrowers. 2019c. 'Vineyard Register Report — New Zealand Winegrowers 2018–2021.' Accessed July 2019. www.nzwine.com/media/12951/vineyard-register-2019_online.pdf.

New Zealand Winegrowers. 2019b. 'Pillars of sustainability: soil, water, air — Air'. Accessed 1 September 2020. www.nzwine.com/media/4339/air-517.pdf.

New Zealand Winegrowers. 2019a. 'Annual Report 2019'. Accessed June 2020. www.nzwine.com/media/15008/nz_annual-report_updated.pdf.

New Zealand Winegrowers. 2020. Biosecurity strategy. Accessed 1 September 2020. www.nzwine.com/en/sustainability/biosecurity/.

Plant and Food Research. 2020. 'Freshfacts — New Zealand horticulture.' Accessed 1 September 2020. www.freshfacts.co.nz/files/freshfacts-2019.pdf.

Plant and Food Research. 2018. 'Freshfacts — New Zealand horticulture.' Accessed 1 September 2020. www.freshfacts.co.nz/files/freshfacts-2018.pdf.

Scandurra, L. 2011. *New Zealand Wine Report 2011*. GAIN Report Number NZ1104. USA: Global Agricultural Information Network. Accessed 1 September 2020. www.globaltrade.net/f/market-research/pdf/New-Zealand/Beverages-Wine-New-Zealand-Wine-Report-2011.html.

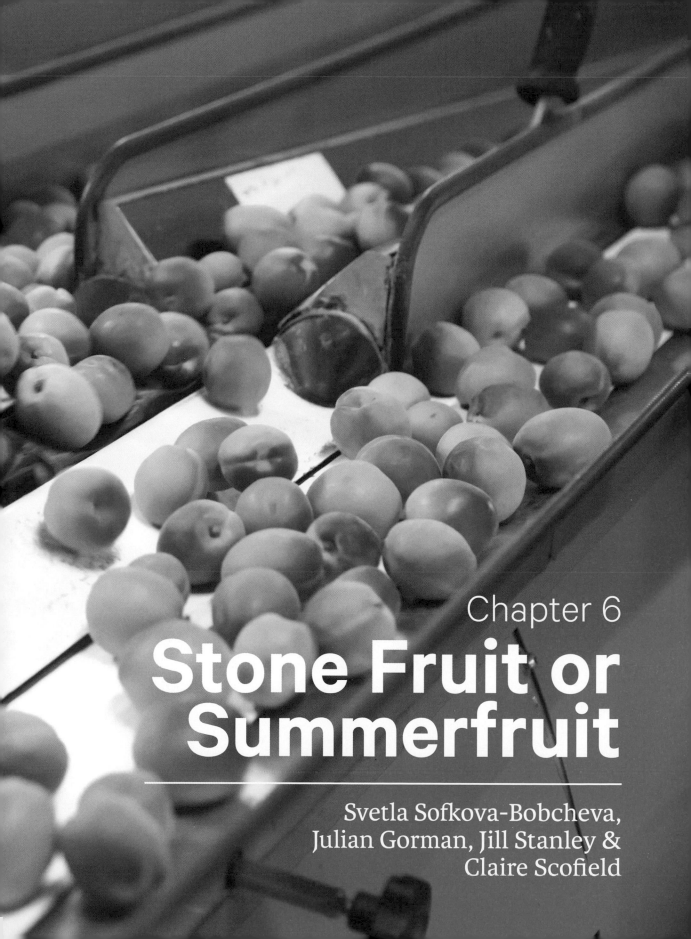

Chapter 6
Stone Fruit or Summerfruit

Svetla Sofkova-Bobcheva,
Julian Gorman, Jill Stanley &
Claire Scofield

Chapter 6

Stone Fruit or Summerfruit

Svetla Sofkova-Bobcheva, Julian Gorman
School of Agriculture and Environment, Massey University

Jill Stanley, Claire Scofield
The New Zealand Institute for Plant and Food Research Limited

Introduction

Summerfruit, also known as stone fruit, comprises apricots, cherries, nectarines, peaches and plums. These are all members of the *Prunus* genus of trees and shrubs. The fruit of *Prunus* species is often defined as *drupe* or stone fruit because the fleshy tissue (mesocarp) surrounding the pip or stone (endocarp) is edible.

The New Zealand summerfruit industry is small in comparison to global production, but vibrant, profitable, sustainable and highly regarded in both the domestic and export markets for the quality of the fruit. In 2018 the global production of apricots was 3.9 million tonnes (MT), cherries 2.6 MT, peaches/nectarines 39.6 MT, and plums and sloes 19.4 MT (FAOSTAT 2018). In comparison, the total crop volume of all New Zealand's summerfruit production in 2017 and 2018 was only 17,690 and 15,541 tonnes, respectively (Summerfruit New Zealand 2018) (Table 6.1).

New Zealand summerfruit industry

In 2020, the industry comprised around 240 growers with a total of 1940 hectares of orchards. Most commercial production of summerfruit takes place in Central Otago (61 per cent) and Hawke's Bay (32 per cent), with small amounts grown in Marlborough and Nelson (Table 6.2).

In Central Otago, the first fruit trees were said to have been planted at Clyde in the 1860s (Buchan et al. 1999). The summerfruit industry started on the back of a decline of alluvial gold mining. The isolation of the district, and the difficulty of transporting a perishable product slowed the expansion of the industry until 1906 when a railway to Alexandra was established (Buchan et al. 1999).

The summerfruit sector is expanding and recently there have been large plantings of cherries in both Central Otago and Hawke's Bay, which are driving export volumes. There has been a small increase in apricot plantings as the result of the recent release of New Zealand-bred varieties.

With the exception of cherries, most summerfruit grown in New Zealand is sold in the domestic market (Figure 6.1).

Climate preferences

The summerfruit industry relies heavily on weather characteristics for production. Most summerfruit can be grown around the country, but climate affects cropping significantly. In warmer parts, apricots, cherries and European plums will not crop very well because of lack

Table 6.1 Sector profile of New Zealand summerfruit in June 2017.

Summerfruit	Growers (no.)	Planted area (ha)	Crop volume (tonnes)	Domestic value 2016/17	Export value (2017) (million $)	Total value (million $)
Apricots	52	318	2567	6.4	5.2	11.6
Cherries	88	645	5025	16.8	71.2	88.0
Nectarines	56	328	4074	17.1	0.2	17.12
Peaches	73	300	3604	13.6	0.7	14.3
Plums	78	217	2420	8.3	0.3	8.6
Total	280	1808	17,690	62.2	77.6	139.6

Source: Plant and Food Research 2017

Table 6.2 Commercial growing area of summerfruit in New Zealand.

	Total summerfruit*	Apricots	Cherries	Nectarines	Peaches	Plums
Hawke's Bay	31%	x	x	x	x	x
Marlborough	10%	x	x			
Nelson		x				
Central Otago	59%	x	x	x	x	x

* 2016 crop survey. (Adapted from Summerfruit New Zealand 2019)

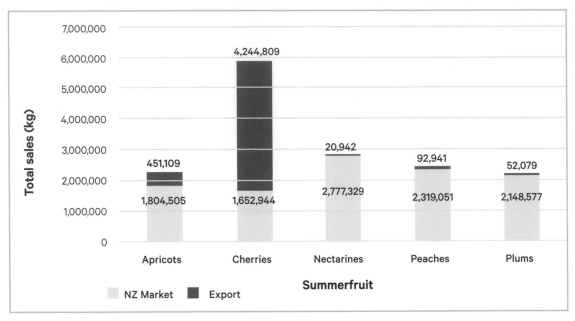

Figure 6.1 New Zealand summerfruit sales, 2017/18 season. (Source: Summerfruit New Zealand 2019)

of winter chilling. In colder regions, late-spring frosts can destroy blossom and young fruit, and low summer temperatures can affect fruit growth.

The main factors influencing summerfruit production, based on international knowledge, include the following.

- Winter chilling is required for bud break and flower development (e.g. cherries) (Palasciano & Gaeta 2017).
- Fruit set can be adversely affected by poor pollination resulting from cool and/or windy conditions affecting bee activity, or by flower death during frost events (e.g. cherries) (Kappell 2010).
- Warm days and cool nights are needed for good flavours. For example, in peaches, a higher soluble solids content is associated with lower minimum temperatures prior to harvest (Johnson et al. 2011).
- Harvest-time must be dry, as rain can mark

fruit, encourage disease, and cause fruit splitting or cracking (e.g. cherry cracking) (Correia et al. 2018).

- Hail storms can seriously damage fruit.

Central Otago is renowned for producing New Zealand's best cherries because of its ideal climate and mineral-rich soil. Central Otago is characterised by hot dry summers and cold dry winters. It lies in the rain shadow of the Southern Alps of New Zealand, and so rainfall is low. Growers rely on bores, irrigation channels and rivers for irrigation. The district is subject to heavy frost (−6 to −9°C) in winter, and light frosts between March and November. Late-spring frosts (late August to late November) can cause severe damage to both blossom and fruit (Buchan et al. 1999).

Producer group

Summerfruit New Zealand (SNZ) is the industry body that represents the interests of New Zealand summerfruit growers, and is the recognised product group for summerfruit under the New Zealand Horticultural Export Authority (NZHEA). The export programme for summerfruit is administered by SNZ, including the market strategy. Export earnings increased from $28 million in 2012/13 to between $70 million and $90 million in recent times. This is predominantly due to cherries, which in 2018/19 generated $69 million in export earnings. The main markets are China and Taiwan — with strong demand for the Chinese New Year holiday in February timed well for the cherry harvest in New Zealand — as well as Vietnam, Thailand and South Korea. Small volumes of apricots, peaches, nectarines and plums are exported to the Pacific and the United States; and more recently, there is interest from Hong Kong in peaches and nectarines (Summerfruit New Zealand 2019).

Commodity levy

A commodity levy applies to the New Zealand summerfruit industry. Such levies are common practice in the primary sector, enabling industries to invest in and solve their own issues. The Commodity Levies Act 1990 allows industries to act on their own behalf and manage their levies. A commodity levy order lasts for six years, and the levy rate needs to be confirmed at an annual general meeting each year.

The levy rate for summerfruit has a cap on what can be collected. For apricots, nectarines, peaches, plums and hybrids thereof, this is up to 1.75 per cent of the price paid at the first point of sale; for cherries up to 1.0 per cent at the first point of sale. The levy money can be used for:

- product development
- research, including market research
- market development
- protection or improvement of plant health
- development and implementation of quality assurance programmes
- education, information or training
- administration of SNZ.

Barriers to summerfruit exports

The quality of the New Zealand fruit in general is exceptional, with good taste and firmness, and no post-harvest defects. However, tariffs have a major impact on the export of summerfruit, with many importing countries (e.g. the European Union, New Caledonia) imposing a tariff. In 2018 cherries to New Caledonia carried a 25 per cent tariff; apricots to the European Union 20 per cent; peaches to French Polynesia 17 per cent; plums to New Caledonia 25 per cent; and nectarines to Fiji 5 per cent. The European Union's tariffs on New Zealand summerfruit are high compared with those for other southern hemisphere exporters. Indeed, some other southern hemisphere countries, such as Chile and South Africa, are not subject to European Union tariffs at all. In 2018, the New Zealand summerfruit industry paid $166,000 in tariffs, which actually represented a 40 per cent reduction on the tariffs paid in 2016. This reduction was largely due to a reduction in apricot exports to the European Union, which carried a 20 per cent tariff (NZHEA n.d.).

Many countries do not allow importation of New Zealand summerfruit, mainly because the biosecurity risks have not been assessed. For example, cherries are the only New Zealand summerfruit that can be exported to China. Discussions have been initiated with countries where New Zealand is keen to have market access, for example apricots to China, and apricots and cherries to Russia. There is usually a waiting list for requests to be considered by the importing country and it can take many years before access is granted. New Zealand will need to demonstrate that risks will be satisfactorily minimised, such as through phytosanitary inspections.

Cherries

The cultivation of cherries is thought to go back to Greek and Roman times, and sweet cherries are thought to have had their origins in the area between the Black and Caspian seas (Patterson 2003). Ancient cherry varieties are genetically similar to many of the commercial varieties grown today, although breeding programmes have produced modern cultivars with improved size, seasonal availability and rain resistance (Patterson 2003).

Cherries are high in the antioxidant anthocyanin, which gives them extra health benefits and appeals to the more health-conscious consumer.

Cherries dominate the summerfruit sector by both volume and value. Central Otago produces most of the cherries grown in New Zealand, with small volumes of pre-Christmas cherries being exported from Marlborough and Hawke's Bay. In the 2017/18 season, 72 per cent of the cherries produced in New Zealand were exported, with an export value of $84.1 million (Figure 6.2). However, in the 2018/19 season the export volume of New Zealand cherries dropped by 2900 tonnes (32 per cent) because of severe weather events at critical times reducing production and export packout rates (MyFarm Investments 2019). In Otago there was a severe frost which caused damage to fruit development, despite

good fruit set. There was also very high rainfall in the Central Otago region in December and January, which affected fruit quality and pack-out rates.

Taiwan is New Zealand's largest market for cherries, with 35 per cent (by value) of the crop being sold to this country in 2018 (Figure 6.3). This volume had reduced from 85 per cent in 2002/03 through industry-initiated market-access projects that targeted new markets to reduce the industry's reliance on Taiwan. The export of cherries to China has increased by 55 per cent since 2016, making it much closer in export value to that for Taiwan. Other key export markets have been more variable, with a $3.4 million decrease (62 per cent) in exports to Korea and a $1.3 million increase (24 per cent) to Vietnam since 2016. There is a fairly diverse range of export destinations for New Zealand cherries — 24 countries imported cherries in 2018, of which nine countries provided export values greater than $1 million (Summerfruit New Zealand 2019).

The production of cherries in New Zealand is usually perfectly timed for the lucrative Chinese New Year. The fruits grown in the Central Otago region are well suited to the Chinese market's requirements as they are firm, fresh, sweet and of high quality. Northern hemisphere growers cannot produce cherries during Chinese New Year so there is little competition, except for Chile, which freights the majority of their fruit by sea. New Zealand's consistently high-quality cherries and the value-added formats (e.g. new packaging and marketing) employed by New Zealand exporters result in New Zealand cherries holding a high price. However, Chile is increasing the volume of cherries being produced for the Chinese New

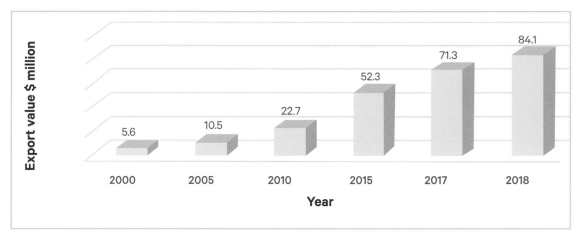

Figure 6.2 Export value ($ million) of cherries from New Zealand between 2000 and 2018. (Source: Fresh Facts 2019)

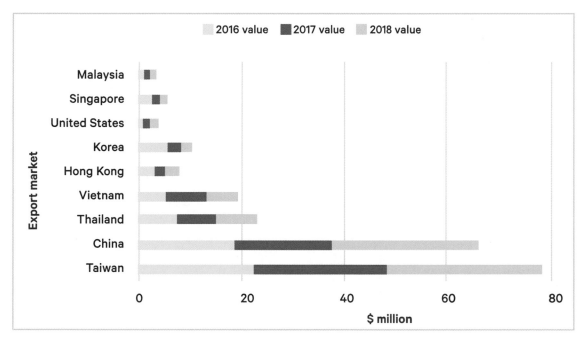

Figure 6.3 Buyers of New Zealand's exported cherries between 2016 and 2018.

Year market, as well as making incremental improvements in quality, and this may pose a threat to future New Zealand exports (MyFarm Investments 2019).

The cherry cultivars grown in New Zealand are derived from the sweet cherry, *Prunus avium*. The sour cherry, *Prunus cerasus*, is mainly used in cooking and is not commonly grown in New Zealand. Dwarf cherries were first researched and grown in New Zealand in the late 1990s by Professor Richard Rowe, but have not yet been taken up by the industry. Cherries are mostly grafted onto 'Colt' rootstocks, which are fairly vigorous; however, some more dwarfing

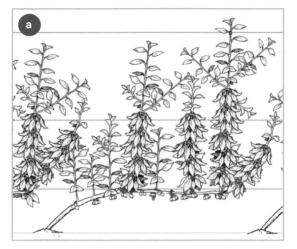 

Figure 6.4 New growing systems for cherries: (a) Upright fruiting offshoots (UFO) system; (b) Narrow-row planar canopies. (Source: (a) Lehnart; (b) Tony Corbett, Plant and Food Research)

rootstocks have recently been introduced into New Zealand and may become more prevalent in the future. New varieties that possess a range of flavours, colours and sizes have been imported under licence agreements.

There have also been many advances in the growing of cherry trees, with new developments featuring much denser planting rates than conventional central-leader or vase-shaped systems. Two new tree-system designs being taken up by many cherry growers in New Zealand involve developing planar canopies. The first system is called the Upright Fruiting Offshoots (UFO) system, in which the upright shoots of single cordon trees are trained into a planar canopy using standard row spacing (Figure 6.4, b; Long et al. 2015). This system improves light distribution within the lower canopy. The second system, developed in New Zealand, is based on two cordons from which upright shoots develop (Figure 6.4, a). In this system the rows are spaced closer together, usually 1.5 or 2 m apart, which increases light interception significantly (Scofield et al. 2020). Prototype 1 trees achieved 3–5 tonnes per hectare (t/ha) in the fourth year (Scofield et al. 2020), whereas an improved prototype on a grower property was achieving over 12 t/ha at the same stage (E. Weaver, pers. comm.).

Birds are a major pest for cherries, and many growers cover trees with bird netting. Cherries are prone to splitting as a result of rain events close to harvest. Plastic covers to protect from rain are becoming more common, although some growers use helicopters or sprayers to dry canopies after rain events. One Central Otago grower has recently erected two automatic retractable covers, which will close for rain and/or cool temperatures (Fresh Plaza 2019b). This not only protects the fruit, but also enables the fruit to mature several weeks earlier than outdoor fruit.

Apricots

Apricots are originally from China but arrived in Europe via Armenia, which is why the scientific name is *Prunus armeniaca*.

Apricots contain a number of potent antioxidants. The apricot is a good source of both vitamin A (from beta-carotene) and vitamin C.

Australia has been New Zealand's main export market for apricots for many years, with 72 per cent of apricot exports being sent there in 2018. There has been a sharp decline in total apricot exports, driven by a 58 per cent drop in exports to Australia as well as substantial decreases in exports to the European Union and United Arab Emirates (Table 6.3). Lower production, poor growing seasons and more complicated market access have all contributed to this (MyFarm Investments 2019).

Clutha Gold and Sundrop are currently the most commonly grown apricot varieties, with Royal Rosa also popular in Hawke's Bay. Apricots are mostly budded on to peach seedling rootstocks. A new grower co-operative, Apricot Co., was set up in 2018 to manage the release,

marketing and licensing of new apricot varieties developed by Plant and Food Research in New Zealand. An interim board, funded by SNZ, was set up as a sub-committee of the organisation to establish the new company. Three new apricot cultivars are yet to be given brand names. Currently known as 'Nzsummer2', 'Nzsummer3' and 'Nzsummer4', they will be named and marketed by Apricot Co. These cultivars, developed in Central Otago, will be grown subject to commercial licence agreements and will be part of a commercial marketing programme.

'Nzsummer2' and 'Nzsummer3' (Figure 6.5) have sweeter, firm flesh with later maturity (mid to late February), good yields and low ethylene production, which results in much slower softening than currently grown export cultivars. These qualities are perfect for extending the apricot season and provide greater storage options and resilience. 'Nzsummer4' matures very early in the season for Central Otago, in early December, yet has good blush and high sweetness for an early-season fruit. It has a short storage life, so is better suited to domestic sales.

Another apricot breeding programme op-

Table 6.3 Five largest importers of New Zealand apricots 2016–2018.

Market	2016		2017		2018	
	Volume	Value (NZ$)	Volume	Value (NZ$)	Volume	Value (NZ$)
Australia	996	4,807,248	1069	4,532,203	445	2,036,062
United States	64	396,163	31	163,123	56	511,686
European Union	169	820,225	66	347,921	32	208,942
United Arab Emirates	95	493,588	21	105,302	4	26,824
New Caledonia	0	0	1	19,376	3	26,374

Figure 6.5 New New Zealand-bred apricot cultivars: 'Nzsummer2' (top) and 'Nzsummer3' (above).

erating up until recently was The Nevis Fruit Company. This breeding company, based near Cromwell, Central Otago, with breeder John McLaren, produced apricot varieties mainly comprised of late-maturing fruit Nevis (Southern Cross, 150, 160, 168, 180).

Australia has long been New Zealand's biggest market, and the apricot breeding programme has responded to customer demand for more brightly coloured fruit with lower acidity and greater sweetness. Summerfruit New Zealand believes that the new apricot varieties have everything that the export markets want, especially the Asian markets (Bus et al. 2016; Stanley, Scofield, and Nixon 2018).

Trial plantings have been set up in Central Otago and Hawke's Bay and both existing and new growers are starting to add these varieties to their orchards. Working closely with Plant and Food Research, the Apricot Co Interim Board (ACIB) was able to supply enough budwood to produce over 20,000 trees for planting in winter 2020. These new cultivars are expected to reinvigorate the industry (Fresh Plaza 2019a). The majority of trees are grown as vase or central leader, although the new narrow-row planar canopies are also showing promise for improving apricot productivity, as well as ease of harvesting, and more new plantings using this system are expected to be seen in the future (Fresh Plaza 2019a).

Peaches and nectarines

Peaches are native to Northwest China, where they were first domesticated and cultivated over 4000 years ago. Nectarine and peach trees are virtually indistinguishable, as are the stones and kernels. The nectarine is a genetic variant of the peach, which has a smooth skin because of a recessive gene stopping the growth of plant hairs on the skin surface.

The first European settlers in New Zealand found groves of wild peaches growing along several North Island rivers. They were known as 'Maori peaches', and it is assumed that they were planted by the crew of the explorer James Cook, or by early nineteenth-century sealing or whaling crews (Dawkins 2008). Peach and nectarine cultivars were among the first fruit trees imported to New Zealand from the Thomas Rivers nursery in England, but in the late nineteenth century locally selected seedlings were found to be better suited to New Zealand's soil and climatic conditions. Favourite peach varieties included the Golden Queen variety (raised by E. Reeve around 1906), which has high yields of yellow-fleshed fruit used for canning, and the Paragon variety (raised by H. R. Wright in about 1903).

There has been some variability in demand for different varieties of peaches and nectarines. In the early 2000s there was a shift away from yellow-fleshed varieties towards white-fleshed varieties, some of which, like the variety Coconut Ice, were bred in New Zealand. Coconut Ice is sweeter and has lower acid content than many other varieties, and was focused on the Asian market. In about 2008 there was a swing back to yellow-fleshed varieties. A wide range of varieties of both peaches and nectarines are currently grown, including flat or doughnut-shaped varieties.

Peaches and nectarines are mainly grown in Hawke's Bay and Central Otago. Figure 6.6 illustrates the main markets for peaches and nectarines between 2016 and 2018.

'Interspecies' denotes the crossing of two prunus species and results in hybrids. There are small volumes of peacherines grown in New Zealand, which are a cross between a Golden Queen peach and a nectarine (Dawkins 2008).

Plums

Plums made up about 13.7 per cent of the summerfruit grown in New Zealand in 2017 (Table 6.1). Only a small percentage of the total plum harvest is exported, with the United States, China, Fiji and New Caledonia representing the main markets between 2016 and 2018 (Figure 6.7).

Many different varieties exist. The European greengage is the most popular plum for canning and making jam. Japanese plums flower earlier and are larger. Burbank, Omega and Black Doris are commonly planted in New Zealand. A new variety of plum bred in New Zealand, Malone, is a high-yielding, late-season plum of large size and with excellent flavour. The fruit store well and have good export potential.

Growing conditions and crop management

Cherries and plums prefer sandy loam or loamy textured soils that are free-draining and fertile and have a neutral pH. They are shallow-rooted and do not grow or produce well in soils that are too wet. It is recommended that soils are ripped to assist drainage and that soils are mounded around the tree to allow water run-off around the root zone (Patterson 2003). A pH range of 6.0 to 6.5 is critical, as the required nutrients are then available to the plant — a lower pH will lock up the nutrients and trace elements that the plant needs to produce fruit (Patterson 2003). The main growing area for cherries is Central Otago, which is hot and dry so irrigation is needed. Under-tree mini-sprinklers is the most common method used, especially in the month before harvest. Overhead sprinklers are also useful, especially for prevention of frost damage. Peaches and nectarines like fertile soils that retain moisture but are well drained.

All summerfruit crops require winter chill to ensure good bud break and flowering in spring. Higher winter chill will mean more flowers, earlier flowering and more-compact flowering. Different varieties have different winter chill requirements. There are several methods of calculating winter chill, the most common being Richardson Chill Hours and hours below 7°C. Summerfruit typically requires between 700 and 1300 hours below 7°C (Hortinfo 2010). Winter temperatures in Central Otago usually satisfy chill requirements easily, whereas some summerfruit cultivars do not perform well every year in Hawke's Bay. Peaches and nectarines are frost-tolerant, but spring frosts can significantly reduce the number of viable flowers.

While most peaches and nectarines, and some cherries, apricots and plums, are self-pollinated, all these crops usually benefit from cross-pollination for improved fruit set and fruit size (McLaren et al. 1995). For example, many of the new apricot cultivars grown in New Zealand are self-fertile. However, two of the most commonly grown cultivars in New Zealand, Sundrop and Clutha Gold are known to be self-incompatible. Orchard design, for example, interplanting different culti-

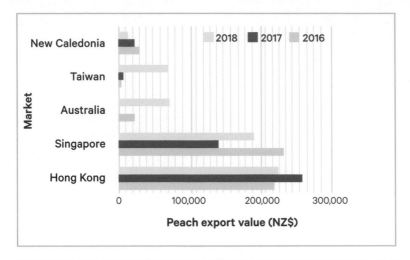

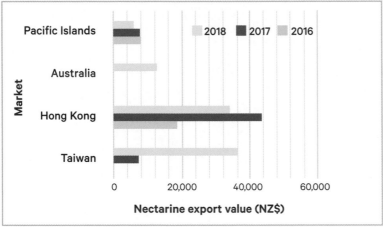

Figure 6.6 Largest markets and export values for New Zealand peaches and nectarines, 2016–2018. (Source: NZHEA n.d.)

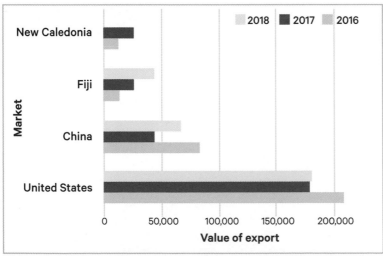

Figure 6.7 Value of plum exports (NZ$), 2016–2018. (Source: NZHEA n.d.)

vars in the orchard, is therefore important.

Low temperatures can affect pollen tube growth and insect activity (Austin et al. 1998; Cerović, Ružić, and Mićić 2000) and frost can damage flowers. Varieties that are self-incompatible require mixed-variety plantings (McLaren and Fraser 1996). All summerfruit crops benefit from the introduction of beehives to orchards over flowering, with 5 hives/ha recommended for apricot orchards (Austin et al. 1996).

There are a number of pests and diseases in summerfruit that need to be managed by growers (Fraser, Morton, and Proebst 2003; Lo, McLaren, and Walker 2000). Pests include aphids, thrips, leaf rollers, mites, cherry slug, mealybugs, Oriental fruit moth and *Carpophilus* beetle. Diseases include brown rot, leaf curl, bladder plum, silver leaf, bacterial spot, bacterial canker and leaf rust. Pests and diseases are managed by monitoring their prevalence and applying agrochemicals as required (Lo, McLaren, and Walker 2000), along with good husbandry practices and the use of beneficial insects such as mite predators, as developed by a SummerGreen Integrated Fruit Production programme (Dawkins 2008).

Biosecurity

The summerfruit industry has always had to keep a close eye on biosecurity. Summerfruit New Zealand has had an active involvement in the recent development of Government Industry Agreements (GIAs) for Readiness and Response. There have been extensive discussions about biosecurity and the need for robust systems to be implemented. In 2017, a deed was submitted to the Ministry for Primary Industries by SNZ and an agreement was signed. Since then, SNZ has become a party to the Fruit Fly Operational Agreement and the Brown Marmorated Stink Bug Operational Agreement. A description of SNZ's most unwanted pests can be found on the organisation's website (www.summerfruitnz.co.nz).

Labour

Labour availability is one of the key factors in continued growth of the summerfruit industry. The availability of seasonal labour is a problem in New Zealand's horticultural industry, with insufficient workers for harvesting, pruning and thinning. The Registered Seasonal Employer (RSE) scheme has been set up to provide seasonal labour from the Pacific Islands. This scheme enables New Zealand growers to continue to output high-quality produce while at the same time providing work and funds for the Pacific (Ministry of Business, Innovation and Employment 2019).

The offshore lockdown in 2020 demonstrated how quickly things can shift in New Zealand's main horticulture export markets. The summerfruit sector has largely escaped the economic downsides of Covid-19 due to the seasonal timing, but post-Covid-19 is the perfect time to review orchard performance and think about employing skilled New Zealanders who can bring new ideas that may boost productivity. Lockdown in New Zealand has seen the development of new sales channels across a breadth of sectors, something that is occurring offshore, too. Some of the questions yet to be answered are: Which new markets can offset some risk and provide new opportunities? How can we cooperate to ensure sufficient and timely airfreight capacity is available for exports? How can we leverage new channels to maximise fresh sales? (Summerfruit New Zealand, 2019a)

'The effects of Covid-19 will be felt across the world for years to come. One thing that has become clear is that everyone needs to eat, and we've been fortunate to have strong support from the Ministry for Primary Industries (MPI) to continue the primary sector operating as essential business to feed both New Zealanders, and other countries,' said Richard Palmer, Chief Executive of Summerfruit NZ. 'Covid-19 has demonstrated our strong connection with others across the horticultural sector, and the importance of the primary sector to New Zealand's economic resilience. We have a unique opportunity to leverage that in collaboration with other like minds.'

References

Austin, P.T., Hewett, E.W., Noiton, D.A., and Plummer, J.A. 1996. 'Cross pollination of "Sundrop" apricot (*Prunus armeniaca* L.) by honeybees.' *New Zealand Journal of Crop and Horticultural Science* 24 (3): 287–94.

Austin, P.T., Hewett, E.W., Noiton, D., and Plummer J.A. 1998. 'Self incompatibility and temperature affect pollen tube growth in "Sundrop" apricot (*Prunus armeniaca* L.).' *Journal of Horticultural Science and Biotechnology* 73 (3): 375–86.

Buchan, Dianne J., Cosslett, C.B., Santorum, A., Webber, D., McLaren, G.F., and Fullerton, R.A. 1999. *Beating the odds: a social and economic analysis of the summerfruit industry in Central Otago*. HortResearch Technical Report 1999/225. Auckland: Horticultural and Food Research Institute of New Zealand.

Bus, V., Nixon, A., Ward, S., and Stanley, J. 2016. 'New Zealand's new apricot selections.' *Produce Plus* 20, Autumn 2016: 27.

Cerović, R., Ružić, Đ., and Mićić, N. 2000. 'Viability of plum ovules at different temperatures.' *Annals of Applied Biology* 137 (1): 53–9.

Correia, S., Schouten, R., Silva, A.P., and Gonçalves, B. 2018. 'Sweet cherry fruit cracking mechanisms and prevention strategies: a review.' *Scientia Horticulturae* 240: 369–77.

Dawkins, M. 2008. 'Stone fruit and the summerfruit industry.' Te Ara — the Encyclopedia of New Zealand. Accessed 27 August 2019. www.TeAra.govt.nz/en/stone-fruit-and-the-summerfruit-industry/print.

Fraser, T., Morton, A., and Proebst, D. 2003. *Organic summerfruit*. Resource guide. Soil and Health Association of NZ Inc., and Bio Dynamic Farming and Gardening Association in NZ Inc. Accessed 27 June 2020. https://biodynamic.org.nz/wp-content/uploads/2014/06/SFRRESGUIDE_LR.pdf.

Fresh Facts. 2000. www.freshfacts.co.nz/files/new-zealand-horticulture-facts-and-figures-2000.pdf.

Fresh Facts. 2005. www.freshfacts.co.nz/files/fresh-facts-2005.pdf .

Fresh Facts. 2010. www.freshfacts.co.nz/files/fresh-facts-2010.pdf.

Fresh Facts. 2017. www.freshfacts.co.nz/files/freshfacts-2017.pdf.

Fresh Facts. 2018. www.freshfacts.co.nz/files/freshfacts-2018.pdf.

Fresh Plaza. 2019a. 'New Zealand apricot growers excited by the release of new varieties.' *Fresh Plaza*, 20 March 2019. Accessed 10 November 2019. www.freshplaza.com/article/9084440/new-zealand-apricot-growers-excited-by-the-release-of-new-varieties/.

Fresh Plaza. 2019b. 'New Zealand cherry grower hosts open house in retractable orchard cover.' *Fresh Plaza*, 12 April 2019. Accessed 26 June 2020. www.freshplaza.com/article/9093335/new-zealand-cherry-grower-hosts-open-house-in-retractable-orchard-cover/.

Hortinfo. 2010. 'Chilling measurement.' March 2006; updated May 2010. Accessed 27 June 2020. www.hortinfo.co.nz/downloads_process_fs.asp?fid=weather%5C%5CFS3-7ChillingMeasurement.pdf.

Johnson, S., Newell, M.J., Reighard, G.L., Robinson, T.L., Taylor, K., and Ward, D. 2011. 'Weather conditions affect fruit weight, harvest date and soluble solids content of "Cresthaven" peaches.' *Acta Horticulturae* 903: 1063–8

Kappel, F. 2010. 'Sweet cherry cultivars vary in their susceptibility to spring frosts.' *Hortscience* 45 (1): 176–7.

Lo, P.L., McLaren, G.F., and Walker, J.T.S. 2000. 'Developments in pest management for integrated fruit production of stonefruit in New Zealand.' *Acta Horticulturae* 525: 93–9.

Long, L., Lang, G., Musacchi, S., and Whiting, M. 2015. 'Cherry training systems.' PNW 667. Oregon State University. Accessed 18 September 2020. www.canr.msu.edu/uploads/resources/pdfs/cherry_training_systems_(e3247).pdf.

McLaren, G.F., and Fraser, J.A. 1996. 'Pollination compatibility of "Sundrop" apricot and its progeny in the "Clutha" series.' *New Zealand Journal of Crop and Horticultural Science* 24 (1): 47–53.

McLaren, G.F., Fraser, J.A. and Grant, J.E. 1995. 'Pollination compatibility of Apricots grown in Central Otago, New Zealand.' *Acta Horticulturae* 384: 385–90.

Milatović, D., Nikolić, D., Radović, A. 2016. 'The effect of temperature on pollen germination and pollen tube growth of apricot cultivars.' *Acta Horticulturae* 1139: 359–362.

Ministry of Business, Innovation and Employment. 2019. 'Recognised Employment Scheme (RES) research.' Accessed 10 November 2019. www.immigration.govt.nz/about-us/research-and-statistics/research-reports/recognised-seasonal-employer-rse-scheme.

MyFarm Investments. August 2019. 'Cherry – market insights.' Accessed 10 November 2019. file:///E:/Massey/Horticulture/Horticulture%20in%20New%20Zealand%20_%20text/Chapter%207%20_%20Soft%20Fruit/cherry-market-insight-august.pdf.

New Zealand Horticultural Export Authority (NZHEA). n.d. '6.7.1. Summerfruit industry profile.' Accessed 10 November 2019. www.hea.co.nz/2012-05-11-03-05-28/summerfruit-trade.

Palasciano, M., and Gaeta, L. 2017. 'Comparison of different models for chilling requirements evaluation of sweet cherry cultivars in a Mediterranean area.' *Acta Horticulturae* 1161: 405–10.

Patterson, M. 2003. *Cherries in Central Otago — feasible or folly? An analysis of traditional and dwarf varieties and methods for the Teviot valley, Central Otago.* Analysis of the cherry industry. For: The Primary Industry Council / Kellogg's Rural Leadership Programme, pp. 33.

Scofield, C., Stanley, J., Schurmann, M., Marshall, R., Breen, K., Tustin, T., and Alavi, M. 2020. 'Light interception and yield of sweet cherry and apricot trees grown as a planar cordon orchard system design.' *Acta Horticulturae* 1281: 213–22.

Stanley, J., Scofield, C., and Nixon, A. 2018. 'New apricot cultivars hit the sweet spot.' *Summerfruit*, July: 18–20.

Statistica. 2017. 'Global fruit production in 2017 by variety (in metric tonnes).' Accessed 10 November 2019. www.statista.com/statistics/264001/worldwide-production-of-fruit-by-variety/.

Summerfruit New Zealand. 2018. 'Sensational summerfruit: a bold plan for growth.' Accessed 10 November 2019.: issuu.com/sumfru12/docs/sensational_summerfruit.

Summerfruit New Zealand. 2019. 'New Zealand summerfruit market data 2017–18.' Accessed 10 November 2019. www.summerfruitnz.co.nz/industry/market-data/market-data-2/.

Summerfruit New Zealand. 2019. Annual Report, www.summerfruitnz.co.nz/publications/annualreport/.

Chapter 7

The Pipfruit Industry

Svetla Sofkova-Bobcheva,
Julian Gorman & Huub Kerckhoffs

Chapter 7

The Pipfruit Industry

Svetla Sofkova-Bobcheva, Julian Gorman
School of Agriculture and Environment, Massey University

Huub Kerckhoffs
Horticultural Sector Policy, Ministry for Primary Industries

Introduction

Significance of the global apple industry

Apples are the third most produced fruit in the world, following bananas and watermelon. In 2017, over 83 million tonnes of apples were produced worldwide, with China producing over half of the total (Statistics Portal 2019).

In terms of the domestic consumption of fresh apples, the largest apple producers worldwide are also among the main consumers. In 2019/2020, the countries with the highest consumption were China, the European Union, Turkey, the United States, India, Russia and Iran. New Zealand was ranked 38th with a domestic consumption of 54,900 million tonnes (MT). Russia remains the largest importer of fresh apples, and China is expected to have an increased import demand for higher-quality apples to a record 100,000 tons (90,700 tonnes) due to its domestic apples being of only fair quality (USDA 2019).

Origin and domestication of the apple

The domesticated apple (*Malus pumila*) originated from the mountains of Central Asia (southern Kazakhstan, Kyrgyzstan, Tajikistan and Xinjiang in China), where its wild ancestor, *Malus sieversii*, can be still found today. Apples have been grown for thousands of years in Asia and Europe and today there are well over 7500 cultivars. The domesticated apple is a

Glossary

controlled-atmosphere (CA) storage Storage method for fruit where a combination of altered atmospheric conditions and reduced temperature allows for prolonged storage with only a slow loss of quality.

count size How many apples there are in a carton. The smaller the fruit size/weight, the more apples in a carton.

dry matter The solid component of fruit — essentially what is left when all the water is removed.

dwarfing rootstock Rootstock that restricts growth and tree size.

ENZA Formally the New Zealand Apple and Pear Marketing Board, ENZA oversaw the export of all New Zealand pipfruit from 1948 to 2001. ENZA is now a brand of Turners and Growers (T&G).

grade standard Fruit quality description for a particular market.

IFP Integrated Fruit Production

kgf (kilogramforce) Measure of the amount of pressure required to pierce a piece of fruit. This gives an indication of fruit crispness and texture. The kgf decreases as fruit ripen.

packout Proportion of the harvested crop that meets export standards.

phytosanitary 'Phyto' means plants and 'sanitary' means hygiene or health. Phytosanitary in the context of pipfruit is the prevention of the introduction and/or spread of quarantine pests.

Pipfruit New Zealand Industry-funded organisation that represents the New Zealand pipfruit industry. Refers to apples and pears, because of the small hard seeds (pips) in the centre of the fruit.

rootstock Part of a plant, often underground, from which new above-ground growth can be produced. Apple rootstocks are selected for vigour control (dwarfing rootstocks) and disease/pest tolerance.

soluble solids Measure of combined sugars and acids in the fruit that can affect taste and eating quality. Measured in units of per cent Brix.

sport (in botany) A sport or bud sport is a part of a plant that shows morphological differences from the rest of the plant. The cause is generally thought to be a chance genetic mutation and is used to develop new cultivars.

starch pattern Measure of the amount of carbohydrate in an apple. This converts to sugar as the apple matures and ripens.

tariff Tax or duty to be paid on particular classes of imported or exported goods.

deciduous tree and grows to a height of about 1.8–4.6 m with many different and desirable characteristics.

Significance of the pipfruit industry in New Zealand

The New Zealand apple and pear industry is a major player on the global stage. New Zealand is the sixth largest exporter in the world, with 390,000 tons (353,800 tonnes) exported in 2018/2019; exports are projected to increase to 405,000 tons (367,400 tonnes), with more shipments being sent to Asian markets (USDA 2019). The increase in output is from trees coming into production from an expanded planting area, alongside re-investing in existing orchards, replacing older varieties and planting higher-density orchards (USDA 2019).

The New Zealand apple and pear industry is very different from most other primary sectors in the country. The industry operates in an unregulated commercial environment, and has created world-leading intellectual property in the form of branded cultivars and proprietary production systems that deliver differentiation and significant extra value over commodity markets. This value creation does not just apply to the produce grown in New Zealand. In order to maximise the value of new branded products, cultivars that were developed in and are owned by New Zealand are being commercialised in the northern hemisphere in partnership with growers in other countries.

The New Zealand pipfruit sector, which comprises apples and pears, is now worth $692 million and is well on its way to becoming a $2-billion industry by 2030 with about 60 per cent of the produce being exported annually to more than 60 countries. New Zealand's pipfruit industry is concentrated in two regions, Nelson and Hawke's Bay, and is dominated by apple production. Over the past decade, apple production has become even more concentrated in these regions, with little change in the area of land devoted to apple-growing (Boniface 2019). Exports of New Zealand apples and pears reached a record value of $870 million for 2019, up $500 million in just eight years. It is notable that 85 per cent of the $500 million increase is value rather than volume — an impressive performance.

By 2018 the New Zealand apple and pear industry was ranked by the *World Apple Review* as the most competitive apple industry in the world for each of the previous four years, based on 23 criteria including production efficiency, industry infrastructure, and financial and market factors. A few key factors have contributed to this success. First, New Zealand has several regions with the perfect climatic conditions for growing apples and pears. Second, the industry has invested heavily in breeding new varieties with properties that match those desired by export markets. Third, the structure of the industry has facilitated proactive approaches to dealing with issues and finding the best ways to be competitive in the global market.

There remain some challenges that need attention, including continued investment in technology, husbandry practice (e.g. disease monitoring and control) and environmental stewardship, but the industry is well set up to progress (Figure 7.1).

Origin and history

Apples and pears have been grown in New Zealand since Europeans first settled in the country. They were initially grown for domestic consumption but pipfruit growers were quick to capitalise

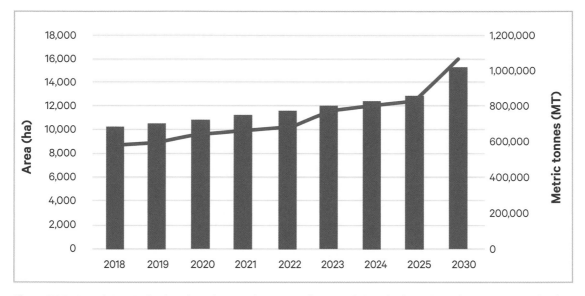

Figure 7.1 Projected New Zealand apple and pear industry growth in area (ha) and volume (MT). (Source: New Zealand Apples and Pears 2020)

on the export potential of their produce.

The missionary Samuel Marsden brought apple and pear trees to New Zealand in 1819. One of his original pear trees is still growing in Kerikeri. The first export of apples from New Zealand was from Christchurch to Chile in 1888, and apples were exported to the United Kingdom in the 1890s.

In the early 1900s the Nelson region became the main pipfruit growing area, producing about two-thirds of New Zealand's apple exports by 1966. Although it remains one of the main pipfruit regions, since the late 1960s there has been considerable expansion of apple-growing in Hawke's Bay. In 2008, over half of the national pipfruit crop was from Hawke's Bay, with one-third from Nelson. The other main apple-growing areas are Central Otago and Waikato.

Apples were the dominant horticultural export crop from New Zealand until the 1980s when the kiwifruit and grape sectors became significant exporters. In the 1960s and 1970s

New Zealand occupied a specific market niche by supplying the northern hemisphere countries with apples during their winter season; however, with the development of controlled-atmosphere storage and other technological advancements this market opportunity has become less significant. New Zealand apples remain competitive in the international market because of their high quality and because new varieties are regularly being produced to suit consumer tastes. Moreover, New Zealand has very efficient production systems and is constantly improving growing practices in order to increase both orchard yields and apple quality.

New Zealand-bred apples

Just over 30 years ago the New Zealand apple industry was facing a difficult situation. Around

60 per cent of the apples exported from the country were Red Delicious and Granny Smith varieties, and these were grown on very widely spaced and large trees, limiting the potential for growth in the industry in terms of land availability, water resources and the labour costs associated with harvesting. Orchardists began changing apple varieties to locally produced Royal Gala and Braeburn cultivars, and orchard production systems changed to semi-intensive and intensive production systems.

New Zealand has led the world in the rapid introduction of new apple cultivars. The New Zealand apple industry has been able to create new apple varieties through its strategic partnerships with key industry players. Established in 2004, PREVAR Limited is a joint venture between Apple & Pear Australia (APAL), Pipfruit New Zealand and Plant and Food Research Rangahau Ahumāra Kai. Commercial cultivars are commercialised through PREVAR Limited and grown under license in New Zealand, Australia, the US, the UK and Europe (Plant and Food Research 2021). New apple varieties include Braeburn, Envy™, Jazz™, Pacific Queen™, Smitten, Sweetie, Rockit, Dazzle and high-colour strains of Royal Gala, Fuji and Cripps Pink/Pink Lady®, which are now premium exports (Table 7.1). As the largest planted variety by area, Royal Gala accounts for 26 per cent and 30 per cent of planted area nationally and in the Hawke's Bay region, respectively, and is ranked as the top variety in export production nationally (NZ Apples and Pears, 2017).

Trademarked apples

The New Zealand pipfruit sector has long had a reputation for innovation. Since 2000, new cultivars have been trademarked with the aim of controlling the number of trees grown and the amount of that specific apple in the marketplace, to keep the return to the growers relatively high. Royalties derived from these new cultivars support the New Zealand breeding programmes. For example, apples from the SciFresh cultivar, sold under the trademark Jazz™, are currently attracting high prices and interest in overseas markets, and the trademark is being developed globally by the exporter ENZA.

Newly developed apple varieties often command a significant price premium compared with the more-traditional apples that are exported. Indeed, many of the varieties that have enjoyed commercial success in recent years have been designed with the increasingly important Asian markets in mind, where consumers often prefer sweeter apples with a higher colour. This is a great example of the New Zealand horticultural sector improving productivity and adding value to exports while still making use of some of New Zealand's key comparative advantages such as favourable climate and access to water.

New apples from old

In the 1920s and 1930s, J.H. Kidd was one of the pioneers of apple breeding in New Zealand. Hoping to combine the heavy cropping and attractive look of American apples with the flavour of English ones, he hand-pollinated different varieties and raised the resulting seedlings. Kidd's first success was from a cross of Delicious with Cox's Orange Pippin, sold as Kidd's Orange Red. Later he crossed Kidd's Orange Red with Golden Delicious to produce Gala.

Table 7.1 Varieties of apple grown in New Zealand.

Variety	Appearance	Harvest time	Origin	Taste/texture
Braeburn	Reddish-orange stripes/blush	Late March to April	1952 chance seedling, Nelson	Sweet, tart, crisp. firm flesh Good storage
Fuji	Soft pink-red blush or stripes, yellow background	Late March to April	1950s cross between Rall's and Red Delicious, Japan	Very sweet with attractive aroma; firm texture and very crisp Good storage
Royal Gala	Red stripes over yellow background	Mid-February to March	Gala mutation, New Zealand	Crisp, sweet taste, firm white flesh
SciRos — Pacific Rose™	Rose pink to bright red; medium to large size	Late March to April	HortResearch breeding programme; Gala x Splendour cross, released 1992	Crisp, sweet and juicy Good storage
SciRed — Pacific Queen™	Large bi-coloured apple with red blush	Early March to early April	Gala x Splendor series developed by HortResearch, 1993	Crisp, sweet and juicy
SciFresh — Jazz™	Bright red	Early March to early April	HortResearch Gala x Braeburn cross; exclusive ENZA variety	Crisp and juicy
Cripps Pink/Pink Lady™	Pale green with pink blush; medium-sized	April	Western Australia, 1984	Crisp, juicy Good keeping qualities
SciEarly — Pacific Beauty™	Bi-coloured with red blush	Early February	HortResearch Gala x Splendour cross	Crisp, juicy, sweet

(modified from New Zealand Apples and Pears)

Pears

Pears are a smaller sector compared with apples. Until the 1980s, pear production in New Zealand relied on overseas cultivars such as Doyenné du Comice from France, William's Bon Chrétien from the United Kingdom, and Packham's Triumph from Australia. This changed as new cultivars from the HortResearch (now Plant and Food Research) breeding programme became available. In the late 1980s a pear grower near Motueka found a natural sport of Doyenné

du Comice with russeted (coarse brown) skin, which was named Taylor's Gold. As the export packout (proportion of fruit qualifying as export grade) from this new cultivar was considerably higher than that of its parent, Taylor's Gold was planted during the 1990s and now comprises 50 per cent of export pears.

Nashi pears

In the mid-1980s there was considerable interest in Asian pears (nashi), which were hailed as the next big horticultural export industry to follow kiwifruit. The first commercial orchards were established in 1984, and the area planted reached 760 hectares (ha) in 1989. Nashi export sales peaked in the mid-1990s. In the early 2000s the area had fallen to about 120 ha, although there has been some recent planting, mainly to cater for the growing Asian population in New Zealand.

Piqa® brand pears

New conventional hybrids of European, Japanese and Chinese pears have been developed at Plant and Food Research and are marketed as Piqa® brand fruits. They have novel flavours, some of which have not been found in pears before, including tropical fruit, tropical pear, melon, coconut and plum as well as recognisable European pear flavours. Piqa® brand fruits are 'ready to eat' at harvest and very well accepted in Asian markets.

The current state of the industry

In 2019 about 10,000 ha were planted with pipfruit and registered for export, with 957 production sites registered to grow apples, 57 export packhouses, 80 exporters, 19.6 million cartons of export apples and pears, and 406,741 tonnes of apple and pears exported (NZ Apples and Pears 2019). For the smaller pear industry, 3594 tonnes of pears were grown in New Zealand in 2017, on 397 ha by 76 growers (Fresh Facts 2018).

The New Zealand apple and pear industry is distributed across specific regions that have comparative growing advantages over other regions of New Zealand and the world. These regions include Hawke's Bay and the East Coast, which grow 70 per cent of the crop; Nelson 23 per cent; Waikato 1 per cent; Wairarapa 1 per cent; Canterbury 2 per cent; and Central Otago 3 per cent.

Marketing and distribution

In the decade to 2018, the volume of New Zealand apple exports has grown 40 per cent, but the value of apple exports has grown much faster — up almost 130 per cent over the same period. Growth has mostly been driven by increased exports to Asia, including strong growth in exports to China. In contrast, apple exports to more-traditional destinations such as the European Union have been fairly stable. Despite the strong growth in exports to Asia, apple export markets remain more diversified than for other New Zealand agricultural products (Boniface 2019).

Apple exports have long been a significant export earner for New Zealand; between 2000 and 2006 they usually generated in excess of $400 million in annual revenues. In 2019, the export value of the pipfruit industry was $692 million with 41 per cent being exported to Asia, the main destination. Other key markets are the United Kingdom, continental Europe and the United States. Knowledge of each market's

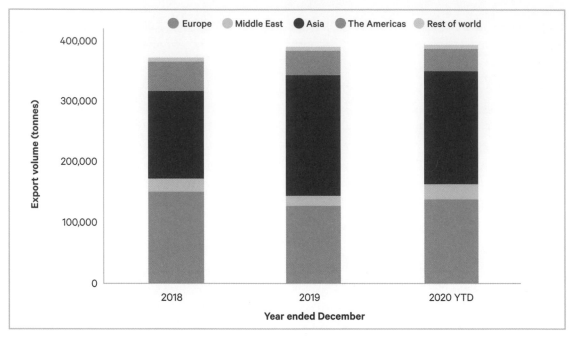

Figure 7.2 New Zealand apple and pear export volumes by destination, 2018–20.
(Source: Statistics New Zealand and MPI–SOPI 2020)

specific requirements in terms of pest and disease quarantine restrictions, fruit size and cultivars is essential (Figure 7.2).

New Zealand pear exports ranged from 2500 to 9300 tonnes between 2001 and 2004. Worldwide, around 18 million tonnes of pears are produced; in 2017 3594 tonnes were grown in New Zealand on 397 ha by 76 growers.

Apple and Pear Marketing Board

Until 2001 all apples were sold by the 'single-desk' seller the New Zealand Apple and Pear Marketing Board, later Enzafruit (ENZ), which was a grower/owner/producer entity similar to Fonterra or Zespri. It was set up to reduce the problem of low prices during gluts and high prices during shortages in the local market, and aimed to bring stability to the export market. The Apple and Pear Marketing Board marketed both export and locally sold fruit.

The pipfruit industry was deregulated in 2001, allowing anyone to sell apples and pears anywhere. Since deregulation, the overall crop volume has changed very little but the number of growers has greatly reduced. There has also been consolidation in the number of packers and exporters. There are currently around 25 significant exporters of New Zealand apples.

New Zealand Apples & Pears Inc.

New Zealand Apples & Pears Inc. (NZAPI) is the representative organisation for the New Zealand apple and pear industry. It also administers, under the Commodity Levies Act 1990, the compulsory apple and pear grower levy, which provides funding for NZAPI to support its activities, which are (1) to support trade and market access; (2) to provide information and knowledge to growers

Table 7.2 Average yield of apples and pears per hectare, 2015–17 (MT).

New Zealand	61.0
South Africa	41.3
Italy	40.0
Chile	39.6
Argentina	36.1
United States	35.6
Brazil	35.5
Netherlands	35.3
France	34.6
Belgium	29.2
Germany	28.9
South Korea	24.4
World Average	**23.4**

highest production per hectare, 48 per cent higher than New Zealand's nearest competitor and 161 per cent higher than the world average (Table 7.2).

Pipfruit are deciduous tree crops that require a period of winter chill. Apple trees need a dormancy period of at least 1200 hours (approximately 50 days) per year below 7.2°C along with a dry, warm summer with intense sunshine (ENZA 2019). During their dormant period, apple and pear trees are reasonably resistant to New Zealand's cold temperatures, frost and snow, especially as winter temperatures are not extreme by world standards.

Frost and hail damage

When buds open in spring, the young leaves and flowers are sensitive to frosts which has, in the past, caused major pipfruit crop losses. To reduce the impact of frosts, growers in Central Otago use water sprinklers, as freezing water into ice crystals releases just enough heat to protect the young flowers, and those in Hawke's Bay use wind machines and/or helicopters to break up the inversion layer created in still, cold conditions.

Hailstorms can cause major production losses by rapidly turning high-earning crops into fruit that can only be used for juice. Selecting sites that are less prone to hail and placing netting over trees is the most proactive way to reduce impacts from hail. Some growers insure their crops against hail damage.

(e.g. Integrated Fruit Production, IFP); (3) to access and develop capability and labour capacity; (4) to develop and commercialise new cultivars; and (5) to represent the industry in global relationships.

Comparative and competitive advantages of New Zealand pipfruit

Climatic factors

New Zealand has an ideal climate for growing temperate fruits such as apples and pears. The ocean that surrounds the country keeps the winters relatively warm and the summers relatively cool compared with competitors that grow in continental climates. This provides a longer growing season and by far the world's

Drought

Drought can seriously reduce both crop and fruit size. Most orchards in New Zealand are irrigated and access to water remains a major constraint for expansion in some areas.

Soils and nutrients

Apples and pears require soils with high fertility, and grow best on well-drained soils with good moisture retention. The high level of nutrients and soil minerals taken up by apple and pear trees needs to be replaced through the application of fertiliser. An apple orchard in New Zealand will yield between 50 and 100 tonnes of apples per hectare. A crop of 70 tonnes on 1 ha of soil removes around 82 kg of potassium, 31 kg of nitrogen, 7 kg of phosphorus and 4 kg each of calcium and magnesium. Some soils, such as those in Hawke's Bay, contain sufficient nitrogen not to need nitrogen based fertiliser; others, such as those in Nelson, are generally poorer in nitrogen and each hectare needs 50 kg of nitrogen, 13 kg of phosphorus and 70 kg of potassium to be added annually (Te Ara 2019).

Low calcium concentrations in soil can cause the disorder bitter pit (small dark spots) in the flesh of the fruit. Pipfruit orchards only deplete about 4 kg of calcium per hectare of soil and this is rectified by foliar spraying of trees with dilute calcium salts between November and harvest time (around February to May).

New Zealand soils are often deficient in magnesium, manganese, boron and zinc, which can affect apple and pear crops. These elements are normally applied using foliar sprays.

Pollination

Apples are self-incompatible: there must be cross-pollination if fruit is to develop. Supplementation of pollinators (honeybees, bumble bees) is often used to increase pollination and fruit set. Apples that do not receive adequate pollination can become malformed as they develop or suffer early fruit drop.

Management practices

Management practices are important to allow growers to best manage their orchards to meet market requirements, from tree establishment to harvest. Managers must understand what the market wants so that they can ensure that the fruit they produce meets their needs and therefore maximises returns.

Some of the key practices used at orchards when growing apples are outlined below.

Variety selection

There are more than 7500 known apple cultivars worldwide (Elzebroek and Wind 2008). The most popular characteristics of commercial apples include a soft but crisp texture, attractive skin colour, absence of russeting (uneven colour and roughness of skin), good storage ability, high yields, disease resistance, common apple shape and desirable flavour.

For many years the New Zealand pipfruit industry was dependent on cultivars from overseas, initially Europe but later North America, Australia and Asia also. Older cultivars were often found to be oddly shaped, russeted, with different textures and colours, to have low yields and be susceptible to disease, along with poor tolerance to storage or transport and often not the desired size.

New Zealand orchardists now produce a premium product because of highly successful domestic breeding programmes, which produce the best-quality pipfruit for each intended market. Apple and pear cultivars vary widely in their time of maturity, skin colour, size, texture, flavour, storage life and susceptibility to storage disorders such as internal browning and softening. Choosing the right variety allows the grower to meet the intended market. For example, Royal Gala was tested after breeding to ensure that it had firm skin and retained its firmness under storage and after transport. Many varieties fail this test, and are scrapped because they will not succeed as an exported apple. Each variety also has its own unique level of soluble solids. While this can be manipulated to a certain extent with reflective mulch, it is still very dependent on the variety. For example, Braeburn going into Europe has a Brix value of 10.5, while for Envy and Pacific Rose it is 15–16.

Rootstock selection

Apple rootstocks

Although apples grow readily from seed, they are usually propagated vegetatively by grafting on to a particular rootstock to ensure that the desirable characteristics of the rootstock, like dwarfing and less vigorous growth habit, are combined with the varietal characteristics. Apple seedlings are heterozygotes, meaning that they tend to produce fruit that is very different from the parent plant. Rootstocks are therefore selected to produce apple trees of a variety sizes that are also winter-hardy and pest-resistant. The desired variety is grafted on to the rootstock to guarantee that the desired fruit is produced.

Clonal (genetically identical) rootstocks are used to promote early cropping, consistent tree size and freedom from most root pests and diseases. This practice also has an effect on harvest time. Trees on M26 dwarfing rootstocks are ready to harvest first, followed a week later by fruit on the convenient semi-dwarfing MM106 rootstock.

In New Zealand, pipfruit have traditionally been grown on large trees on semi-vigorous clonal rootstocks. These rootstocks were originally from the United Kingdom and have been found to be very sensitive to infestations of woolly apple aphid (*Eriosoma lanigerum*). In response to this, a breeding programme in England in the 1920s produced the Merton (M) and Malling Merton (MM) apple rootstocks. These are resistant to infestations of woolly apple aphid and became the mainstay of New Zealand production for many years, particularly the rootstocks M.793 and MM.106.

During the 1970s and 1980s apple-growers changed from producing multi-leader vase-shaped trees to centre-leader cone-shaped trees, with one central trunk and distinct tiers of branches. These begin cropping earlier than multi-leader trees, and picking and tree management and pest control are easier.

In addition, dwarf rootstocks produce smaller trees, which can be planted more intensively and are easier to manage. Growers began to use these rootstocks, particularly M9, in the 1990s. The trend continued into the 2000s, with plant densities of 1250 to sometimes 3000 trees per hectare on dwarf rootstocks. The rootstock M9 is sensitive to woolly apple aphid, but the pest's effect on the industry had already been reduced by the deliberate introduction in 1921 of the tiny wasp *Aphelinus mali*, which destroys the aphid.

Intensive planting with dwarf rootstocks allows earlier production in terms of fewer years to full productivity (Table 7.3), improved fruit quality and easier tree management. Much tree husbandry can be done from the ground. Establishing an orchard is costly, but early production of high-quality fruit means that the orchard can reach break-even point in five years or less, particularly with new apple cultivars that fetch high prices.

Table 7.3 Comparison of two apple production systems based on different varieties and rootstocks.

	Production system 1	Production system 2
Varieties	MM793, MM106, MARC rootstocks	M9 and sports, M7 and sports, M26, CG202, CG210 'Restricting' dwarf rootstocks
Size	Large, up to 6 m	Smaller, 3–3.5 m
Spacing	4 m in row, 5 m between rows	0.8–1.1 m in row, 3.1 m between rows
Density	about 666 trees/ha	2000+ trees/ha
Years to cash flow	5	2
Years to full productivity	7	4
Life	20+ years	Earliest plantings now 15 years

Pear rootstocks

Pear trees are normally grafted on to clonal quince rootstocks, which are easy to propagate vegetatively and allow the pear-grower to control the size of trees and have them fruit early. However, not all pear cultivars are graft-compatible with quince, so a stem of a graft-compatible scion cultivar is grafted between the rootstock and the desired cultivar.

Structure of apple orchards

Pruning and training

Pruning and training of trees is extremely important in pipfruit orchards, and there has been a continuous drive to innovate regarding the structure of orchards to further optimise yield and quality.

Winter pruning removes one-third of the flowers before the tree starts to grow again in spring. This helps the tree to produce fruit of the ideal size. It also removes older wood so that more new, vigorous wood can fruit. More light is also able to penetrate the canopy, increasing the soluble solids.

The method of pruning an apple tree, and how much new and old growth the tree has, both have an effect on fruiting. Trees with a lot of newer growth (such as one-year-old wood) will fruit earlier than trees with a lot of older fruiting wood. It is recommended that growers harvest 10 days earlier for one-year-old wood. New plantings on dwarfing rootstocks will also fruit very heavily compared with very old trees in older blocks (Figure 7.3).

ReTain®

ReTain® is a chemical that is applied to an apple tree to retain the fruit for longer. This allows the fruit to grow a little larger without over-ripening. For example, fruit that is ready to harvest (shown by the starch patterns having reduced) would move count size from 130 to 110 if the tree is sprayed with ReTain®. This allows the grower to sell to markets that want larger fruit. Spraying with ReTain® also allows a tree to be harvested with fewer picks over a shorter period of time.

Thinning

Chemical, hand and mechanical thinning reduce the amount of fruit on the tree. The fruit that is left can therefore grow to a marketable size (not too big and not too small). For example, a conventional tree on MM106 rootstock can produce around 2500 flowers and it would be desirable to reduce these down to allow 400 fruit to be harvested. Some flowers will be dropped naturally while others need to be removed. Spraying with chemicals such as sulphur will burn off flowers. Thinning by hand provides a good opportunity to also remove deformed fruit that would later get graded out anyway. It also allows the grower to thin out bunches to stop fruit from rubbing together (which leaves blemishes on the fruit).

Frost protection

Frost events occur in late winter and into early spring. Frosts can kill off flowers, reducing the amount of fruit produced by the tree and consequently producing larger fruit. This can be catastrophic when growers have already thinned the tree to the correct level before the frost hits. Frosts that occur later in the growing season can also damage the skin of the fruit and as the fruit continues to grow, the blemish also grows. This lowers the grade of the fruit and therefore its value. Frost pots, helicopters or wind machines are used to prevent cold air settling (and a frost occurring) by mixing layers of air.

Reflective mulch

Reflective mulch is a fabric that is laid down between rows of apple trees, allowing light to bounce back up into the lower branches that do not get as much sun as the higher branches. This helps to colour up the fruit and ensures that the fruit on the lower part of the tree reach the required level of soluble solids before harvest. It also allows the fruit to be harvested over a shorter period of time; usually the trees need around four picks to harvest all the fruit.

Pests and diseases

Fire blight

Because apples and pears were introduced to New Zealand both as whole trees and as graftwood, many pests and diseases were introduced at the same time. The most serious bacterial disease, fire blight (*Erwinia amylovora*), was introduced in graftwood from the United States after World War I. This resulted in a ban on the export of New Zealand pipfruit to Australia, which has remained free of the disease; this ban was not lifted until 2011, as discussed at the end of the chapter.

Fungal diseases

Two major diseases of apple trees are black spot (apple scab, caused by *Venturia inaequalis*) and powdery mildew (*Podosphaera leucotricha*). The former is more prevalent in wetter areas and the latter in drier regions. Black spot can cause major losses of fruit and leaves if not controlled with fungicide sprays. A major objective for New Zealand breeding programmes is to produce new cultivars that are resistant to these diseases.

Pests

Major pests of pipfruit include leafroller caterpillar (Tortricidae moth family), codling moth (*Cydia pomonella*), woolly apple aphid

(*Eriosoma lanigerum*), leafcurling midge (*Dasineura mali*) and European red mite (*Panonychus ulmi*). Although these pests can be controlled by insecticidal sprays, much scientific work has gone into understanding their specific biology and the role of predators and parasites in their control. Spray programmes in the 1980s relied heavily on broad-spectrum insecticides that were toxic to a wide range of insects, both pest and predator. Pests became resistant to some of these materials, which has led to a greater interest in biological control.

Reducing chemical residues

By the early 1990s, consumers had become concerned about the regular use of broad-spectrum chemicals in orchards. In response, the New Zealand pipfruit industry developed the Integrated Fruit Production (IFP) programme, which relies on ecologically safer methods of pest control. Trees are sprayed only in response to pests or diseases rather than a spraying programme being dictated by the calendar. The chemicals used are primarily toxic to the specific pest so that predators and parasites of the pest remain unharmed.

An example of the change in approach is that of the leaf-sucking European red mite, which became a serious pest in New Zealand apple orchards when it developed resistance to some spray chemicals. Other mites that ate the pest were also killed by the sprays. Growers now monitor populations of the red mite and its predators before deciding whether to spray. They often rely totally on the predatory mites to limit pest numbers and damage.

Integrated Fruit Production (IFP) programme

The IFP programme was adopted by the pipfruit industry in 1996, and by 2001 all export fruit was being exported under IFP. The programme was developed in response to strong demand from the export markets — in 2002 there were about 90 exporters of various sizes and the phytosanitary requirements of the various export markets required high standards. New Zealand's IFP guidelines were developed from those of the International Organisation for Biological Control (IOBC). The international guidelines were matched carefully with local production conditions, and once implemented resulted in a 95 per cent reduction in organophosphate pesticides and a 50 per cent reduction in overall insecticide use (Wiltshire 2003). The programme is continually being updated as further work is carried out on pest and disease biology. In 2007 the pipfruit industry began trialling a zero chemical residue programme.

The Apple Futures project, a government and industry funded project aimed at helping the pipfruit industry tackle the growing demand in international markets for fruit that has no detectable pesticide residues, is the next significant step down the pathway towards the production of safe fruit in a sustainable way. The programme has been very successful at underpinning New Zealand's strong position in the world market. This was achieved by meeting the regulatory phytosanitary market-access challenges of the high-paying Asian markets and also exceeding the minimum residue level (MRL) requirements and Good Agricultural Practice (GAP) requirements of the high-paying European markets.

These innovations have helped to maintain access to the top North American and European retailers and have clearly demonstrated that the developing markets of Asia, the Middle East and North Africa are where the industry's future

growth lies. New Zealand apples won't be the cheapest on offer, so success hinges on being able to differentiate the product and convince consumers to pay for a product that is of superior quality.

Organic pipfruit

Organic pipfruit growers do not use synthetically produced pesticides and fertilisers, but instead rely on copper- or sulphur-based fungicides, and organic fertilisers. Most organic pipfruit production occurs in Hawke's Bay, as the drier climate makes the control of diseases such as apple scab and summer fruit rots easier. The rich soils also help with tree nutrition. The costs of production are higher, as it is more labour-intensive — but organic fruit sells for a higher price. Some growers have adopted organic methods for philosophical reasons, but for others it is a purely economic decision. In 2018 the average annual harvest of organic apples in New Zealand was 32,500 MT with 72 per cent of these being exported to Asian and US markets. Bostock New Zealand is the leading organic grower and with 700 hectares of Bio-Gro certified organic orchards produces 85 per cent of New Zealand's organic apples.

Time of harvest

The decision on when to harvest fruit is based on maturity. If apples are harvested when immature, they will not have the right taste and aroma; if left on the tree too long, they soften and may develop a greasy skin. Orchardists carefully monitor fruit maturity, looking at skin colour, flesh firmness, soluble solids, and starch breakdown patterns (as shown by the starch–iodine reaction on a transverse section of the fruit). Each cultivar has its own optimum

values for these factors, which predict the best cool-storage life.

Picking and packing

The apple-picking season lasts several months as each cultivar becomes ready for harvest. Export harvesting may begin with Cox's Orange Pippin in February and finish in April with Granny Smith. Most cultivars are selectively harvested in three to four picks over a two- to three-week period, with each pick removing the most optimally mature fruit. European pears are normally harvested in a once-over pick.

Apples and pears are picked by hand, usually into bags slung across the picker's chest. The fruit is emptied into larger wooden bulk bins, holding around 400 kg, which are moved to a cool-store or a grading shed. Apples are usually dumped in water and floated off to be cleaned and graded. They are graded for freedom from visible defects, usually by human eye on a grading table, then pass under electronic colour-sorters and weigh-cells before passing onto separate lanes for packing. Fruit are normally packed into 18 kg cardboard boxes, with each layer of fruit being placed on a fibre tray for protection during transport.

Pears are not normally dumped in water, as they do not float. Instead, they are removed from the smaller pear bins by hand and placed on the grader. Pears are usually packed in small, two-layer cartons which hold around 7.5 kg.

Storage and transport

All horticultural products are seasonal and many types of fruit are highly perishable. Apples store well and so are available over a longer period of time. The key export period for New Zealand apples is March to September and almost all

New Zealand export apples are exported to the northern hemisphere.

The preservation of fruit quality during storage and transportation requires the manipulation of temperature, oxygen and carbon dioxide. This is usually achieved through controlled-atmosphere (CA) storage, where oxygen is decreased in the air within a cool-store and the concentration of other gases (such as nitrogen and carbon dioxide) is adjusted to prevent the fruit ripening. Additionally, the fruit is stored at the very low temperature of 0.5°C. For most producers only a small part of the crop can be stored this way due to the limited facilities available, which forces them to pack all of their apples and sell them as quickly as possible. Other producers, such as Apollo Apples with 30,000 m² of cool-storage, have gained a significant competitive advantage by delivering fruit every week or fortnight over many months to customers who want certainty of supply.

As well as temperature and atmosphere control, storage life can be extended by removing ethylene from the cool-store. Ethylene is a gas that is associated with ripening of fruit.

A major advantage of the various advances in storage technology is that high-quality fruit can be delivered at regular intervals all year round. The ability to preserve and transport fruit over long distances without it spoiling in the bulk holds of large ships has reduced costs considerably. However, these technologies have also allowed northern-hemisphere orchardists to store and use their own fruit during winter instead of being dependent on a southern-hemisphere supply.

Sustainability

Pipfruit production inevitably has an impact on the natural environment and on the society in which it takes place. Apple orchards are generally classified as intensive forms of land use, traditionally needing high inputs of fertilisers, mechanical energy, labour and agrichemicals. However, the industry has become increasingly committed in recent years to ensuring that the apples are safe, production practices are sustainable, and the use of agrichemicals is minimised. Close monitoring in orchards and packhouses using traceability systems provides further quality assurance.

Environmental sustainability is a key focus for New Zealand's pipfruit industry. Plant and Food Research in collaboration with the industry has developed many of the tools required to help limit the impact of pipfruit-industry production systems on the environment. This ensures that New Zealand continues to exceed the world's most stringent sustainability requirements.

GLOBALGAP and GRASP (GLOBALGAP Risk Assessment on Social Practice)

GLOBALGAP began in 1997 as EurepGAP, an initiative by retailers belonging to the Euro-Retailer Produce Working Group. British retailers working together with supermarkets in continental Europe became aware of consumers' growing concerns regarding product safety, environmental impact and the health, safety and welfare of workers and animals. Their solution was to harmonise their own standards and procedures and develop an independent certification system for Good Agricultural Practice (GAP), defined by the Food and

Agriculture Organization of the United Nations (FAO) as a 'collection of principles to apply for on-farm production and post-production processes, resulting in safe and healthy food and non-food agricultural products, while taking into account economic, social and environmental sustainability.' GLOBALGAP today is the world's leading farm assurance program, translating consumer requirements into GAP in more than 135 countries. The New Zealand apple and pear industry was among the earliest adopters of GLOBALGAP, with many businesses having been certified since 1997.

Looking after workers and providing them with a positive employment environment is a fundamental obligation for employers. A positive working environment has a significant impact on both the quality of the product and the effective operation of the production facilities. The value and benefits of well-trained, -encouraged and -protected workers can be seen all along the supply chain. Corporate social responsibility is an increasingly important issue in global food supply chains. Stakeholders increasingly need to demonstrate that their products are produced in line with internationally agreed labour requirements as well as relevant legislation.

Recognised Seasonal Employer (RSE) scheme

The New Zealand pipfruit industry often suffers from a shortage of local workers. The seasonal nature of employment makes these industries difficult to staff during periods of peak demand, and the nature of the product and the quality requirements means that just-in-time production is the only profitable form of production; there are often very narrow tolerances for har-vesting or other tasks. The Recognised Seasonal Employer (RSE) work policy, developed by the Ministry of Business, Innovation and Employment (MBIE), allows for the temporary entry of workers from overseas to plant, maintain, harvest and pack fruit at peak times.

The RSE scheme has been recognised as the world's leading migratory labour scheme, having been described by the International Labour Organization (ILO) and the World Bank as a best-practice managed circular migration scheme. The ILO's good practices database states that 'The comprehensive approach of the RSE scheme towards filling labour shortages in the horticulture and viticulture industries in New Zealand and the system of checks to ensure that the migration process is orderly, fair, and circular could serve as a model for other destination countries.' The United Nations have also put it forward as a model for other countries to follow. The RSE scheme has provided employers with suitable and sustainable volumes of labour with which to pick, pack and maintain their crops with world-leading quality. Pacific Island nations have found the scheme to be hugely beneficial through delivering transformational changes to the communities involved. Mean earnings per worker are around $18,000 per annum with overall RSE earnings of $250 million per year. The scheme was affected in 2020 by the Covid-19 pandemic, with the New Zealand government coordinating with NZAPI to fascilitate border entrance for workers from these countries under the new restrictions.

Robotics and automation

There is much talk about automation in the New Zealand apple and pear sector. We have seen technological advances in the packhouse quickly removing the need for graders. Automated palletisers and pallet stackers are common and tray-filling machines are now being installed commercially. Price look-up (PLU) labels have been used since 1990 to make checkout at the supermarket and inventory control easier, faster and more accurate. A compostable label has recently been introduced and is being trialled in some countries. In a further step, machines have recently been developed that use lasers to mark fruit directly with a code without damaging it on the inside. The introduction of this technology removes previous problems such as stickers that are hard to remove from fruit. The fully automated packhouse ('lights out') may be just around the corner.

The orchard, however, presents a tougher challenge. Key to automation in the orchard is a two-dimensional canopy that provides easier access to the fruit for any machine harvester. Growers are planting around 600 ha each year, comprising around 300 ha of expansion and another 300 ha of redevelopment. Much of this development involves two-dimensional Future Orchard Production Systems (FOPS) type systems. These have the potential to be more easily harvested using the robotic systems that are currently being tested both in New Zealand and elsewhere in the world (Figure 7.4).

Technical breakthroughs will continue to influence horticulture tasks. One significant gain provided by robotics and automation relates to the collection of data. For example, a system such as the Ladybird can scan or photograph an orchard, individually identifying each tree, giving it a GPS location and measuring changes in multi-dimensions. Currently most of advances are in computing capacity, where algorithms access and process vast quantities of data, improving our ability to implement and enhance current technologies. In future we will see the development of commercial-scale autonomous technologies or robotics. The horticulture sector in New Zealand is already using new commercial systems in the data-processing space but robotic harvesting systems are also being developed and tested in our orchards. These current and developing technologies will require unique competencies and a skills and education system that is nimbler and more responsive than before.

Market requirements

Each market to which New Zealand exports apples has specific requirements. These requirements affect decision-making at the start of the season as pre-harvest management needs to follow the specific requirements of the market that the grower is aiming at in order to gain access to that market (Table 7.4).

It is sensible for New Zealand to export a range of different apple varieties to a variety of markets. The bulk of apple exports are dominantly fresh fruit of different varieties, sizes and grade standards. Approximately 6 per cent of exports are from organic production systems, with all other fruit produced under a sustainable integrated fruit production (IFP) programme.

Apple attributes

Apple attributes refer to specific quality features of the apple, such as variety, count size, colour, soluble solids, flesh firmness, quarantine pests, chemical residues and grade standard.

Each market has different requirements for

Figure 7.3 Older free-standing vase-shaped trees are increasingly being replaced by (a) centre-leader (666–2500 trees/ha), (b) V-shaped trellis (2000–2500 trees/ha or (c) planar cordon vertical design or 2D-trellis (2500–4000 trees/ha) plantings. Following pruning, each upright mostly has only short shoots (spurs). This, along with the vertical orientation and spacing of the uprights, allows light to penetrate into the bottom part of the tree.

apple imports, and apple producers aim to supply the market that pays the highest return. Some markets insist on strict biosecurity measures and need assurance that fruit does not have a particular unwanted pest or disease, or require that fruit must not have chemical residues above certain limits (the minimum residue level, MRL). These regulatory requirements are enforced through government regulation in the importing country. Customers within these markets often demand specific 'commercial grade-standard requirements' such as colour, variety, size and firmness.

It is more difficult to grow fruit to meet the requirements of the higher-paying, more demanding markets, but motivated growers who are keen to make a profit take great care to ensure that their apples meet the needs of the market. For example, successful growers manipulate their management practices at the orchard to get the quality and count size that maximises their returns. A considerable amount of time and money is spent getting fruit to meet export standards and phytosanitary restrictions, but the increased returns make the effort worthwhile. Not all fruit will pass the regulatory requirements of the export markets and is then often sold on the New Zealand domestic market.

Table 7.4 The changing face of market access requirements.

Late 1980s and early 1990s	Mid- to late 1990s	Circa 2000	Today
Phytosanitary	Phytosanitary	Phytosanitary	Phytosanitary
	Regulatory residues	Regulatory residues	Regulatory residues
	Official grade standards	Official grade standards	Official grade standards
		GAP (customer service)	Commercial residues
		Food safety	Genetic and customer GAP
		Labelling and traceability	Variety — colour, size, firmness, soluble solids
		Customer programmes	Labelling and traceability
			Customer programmes
			Food safety
			Ethical trading
			Sustainability
			Trade security

Time of year

Apple markets, such as those in Europe, the United States and Taiwan, have specific timing requirements for two main reasons. First, countries in the northern hemisphere have traditionally bought New Zealand apples during their winters (our summers), as apples are a seasonal fruit that can't be produced year-round. This is called 'counter-seasonal production' and ensures that these countries have apples on their supermarket shelves all year. Second, to protect the apple producers in the northern hemisphere, countries impose tariffs on imported fruit from a predetermined date to make sure that southern-hemisphere fruit does not compete with fruit produced domestically in the northern hemisphere.

New Zealand's supply season is dictated by seasonal weather patterns that control bud burst and fruit production. Growers can manipulate the trees to flower within a small window but seasonal weather patterns are the major influence. In recent years the significant advancements in fruit storage methods have reduced the demand for New Zealand apples where countries produce enough fruit of their own to either sell fresh or store to sell later. This has put New Zealand's strong position in the market at risk, as apples can be stored for up to 12 months. New Zealand has, however, continued to build its strong reputation for producing high-quality and safe (pest and chemical-free) apples.

Quantity

New Zealand has a strong reputation built on consistency of supply. If New Zealand apple exporters cannot supply the required market volumes consistently from year to year, they will turn to other countries (such as Chile) to satisfy the demand. Chile is known to be inconsistent in its supply internationally; climatic and quality challenges, along with shifting market preferences, mean that the country fails to build strong supply relationships of its own.

To gain access into a market, apple producers often have to supply many varieties of apples at sufficient volumes. Apple sales follow a seasonal pattern, with early varieties harvested and exported to be placed on retailer shelves first. Retailers will often change varieties as the sales season progresses. Retailers demand that their suppliers send them a whole-season programme of varieties of certain volumes, sizes and colours. Exporters need to produce sufficient volume and choice of varieties with the right quality and size to supply to their retailer.

Quality

The quality of New Zealand apples is one of the country's key points of difference with other apple-producing countries. Markets have specifications to meet regarding various quality standards (Class 1, Class 2, etc.) and phytosanitary requirements. All apples are graded following harvesting to separate the fruit by quality. Grading of apples is based on size, colour and defects such as bruising or blemishes on the skin. Export-quality fruit has very few imperfections so that it is visually appealing to consumers. Apples for the domestic market might have some imperfections that mean that they are not of sufficient quality to demand the higher export prices.

New Zealand has placed itself in a quality category of its own because of the very high standards it consistently meets, including low chemical residues in the fruit. Producers do not just meet the minimum standards set internationally; they have gone beyond these to produce fruit with virtually untraceable amounts of chemicals. This ensures that New Zealand apples are in high demand in markets where safety and product integrity are prioritised by consumers. The standards that New Zealand apples meet helps to ensure that they receive the highest payments possible, particularly in the European Union, where they often comply well below the environmental regulations.

While New Zealand is relatively free of the pests and diseases that can plague other apple-producing nations, there are still a few that other countries wish to exclude, so phytosanitary programmes must be in place to mitigate the risk of exporting these. These programmes typically involve both chemical sprays and biological controls. Any incursions will immediately trigger border closures by importing counties and thus have a major and direct economic impact on the New Zealand industry.

Significant effort and funding is also put into biosecurity to keep New Zealand free of any unwanted pests and diseases from other countries, like the Queensland fruit fly and brown marmorated stink bug (BMSB).

Packhouses expend major effort to ensure that the highest number of harvested fruits get exported. Many factors can influence this, such as frosts, strong winds, hail or insect pest damage. Some growers adopt a very different approach, such as targeting the processed-food or juice markets that have less-demanding quality requirements for apples. If growers keeps their production costs sufficiently low (i.e. not spending too much time or money trying to achieve a specific size, colour, grade or phytosanitary specification), then they are able to gain sufficiently high returns to still be profitable. Much of the fruit rejected for export is used for juice or other processing, however, so there are limitations to the number of growers who can grow solely for the juice market.

Consistency

Apples are graded in modern packhouses that use sophisticated, cutting-edge technology to sort the fruits into their correct count sizes, colour bands and blemish tolerances. This ensures that each market is provided with the size, colour and quality of fruit that it needs or is prepared to pay for. Another consideration is the quality from year to year; consumers want and expect the same quality each year.

A third issue relating to consistency is the volume supplied to market. Each market has similar demands from year to year, as their population and needs do not fluctuate greatly. If similar volumes are supplied each year, the markets' demands are satisfied and they continue to demand fruit from New Zealand. Some variation in demand is, however, inevitable. Markets might demand less fruit if their own domestic production is high or if higher prices encouraged domestic producers to ramp up production. If the weather had been particularly poor or the prices over past seasons were low, local producers might produce less fruit and so there would be more demand for imported apples.

Future challenges and opportunities

The New Zealand apple and pear industry has grown faster than any other export industry in New Zealand. In 2012, New Zealand industries

were challenged to double exports by 2025 — and the apple and pear industry had achieved this by 2016. In 2019, with a forecasted export value of over $800 million, the industry has its sights well beyond the $1 billion mark.

Seasonal and permanent staff needs projection

The NZAPI conducted extensive surveying of its members in 2018. Those surveys indicated that for every hectare of orchard planted, the pipfruit industry needs to employ 0.17 full-time permanent (FTE) orchard workers. For every 1000 MT produced, the industry needs to employ 2.36 post-harvest permanent workers. And for every 1000 MT packed, the industry needs to employ 1.25 permanent workers in corporate head-office jobs (NZAPI 2020).

In order to meet a target of $2 billion by 2030, the apple and pear industry will need an additional 876 permanent orchard workers and 1322 post-harvest and corporate services workers. The total permanent workforce will increase by 2198 workers, or 60 per cent, by 2030.

Growing capability

Over the past 20 years there has been a hollowing-out of horticultural capacity at the two main primary-industry universities of Lincoln and Massey. With growth in horticulture taking off and the long-term economic fundamentals looking strong, it is not surprising that Massey University is refocusing on strengthening its horticulture teaching with new Bachelor of Horticultural Science (BHortSci), Bachelor of Horticulture (BHort) and Master of Horticulture (MHort) degrees being developed. Fruit businesses are becoming larger, more sophisticated and vertically integrated (from growers to packhouses to exporters). These new corporate businesses require new skills and capability to allow them to grow. Due to the increased sophistication of pipfruit businesses and global fruit trade complexity it is not surprising that horticultural employers are increasingly employing university graduates to meet their competency demands. Additional roles include sales staff, engineers, business analysts and ICT (information and communications technology) professionals. Formal qualifications will be required in fields such as horticulture, commerce, sciences and social sciences.

Sustainability

The environment in which the New Zealand apple and pear industry is operating has changed fundamentally in recent years. Global trade drivers are changing, and countries are questioning the sustainability of their imported produce. New Zealand has been an enthusiastic participant in the development of a rules-based platform through its active participation in the World Trade Organization and has been a strong advocate for free-trade agreements (FTAs).

Market trends

The New Zealand apple industry is currently focusing on Asia and the Middle East, where economies are growing and consumers are still discovering apples. Population growth is strong in these countries, and the safe and healthy apple that New Zealand is renowned for is perfect for affluent consumers.

Older and more traditional markets such as Europe and the United Kingdom tend to have ageing populations and low or falling populations. The 2008 Global Financial Crisis (GFC) hit these markets hard, and their economies have stagnated. The New Zealand apple industry is aiming to maintain volumes into these markets.

The 2020 apple and pear export season indicated increases in export volumes and prices compared to the previous season despite on-shore and off-shore disruptions from responses to the Covid-19 pandemic. Consumer demand for fresh fruit remained strong despite Covid-19-related disruptions, and this is expected to continue. However, seasonal labour supply concerns and logistics constraints may make it challenging for the upcoming harvests to reach their full potential. Apple exports to Asia were impacted by Covid-19- induced disruptions to shipping schedules and market outlets. Market demand from Europe was strong, driven by a reduced domestic apple crop in 2019, and by supermarkets, which remained open during 2020, the main retail outlet for apples (SOPI 2020).

Export constraints and challenges

Australia

From 1921, New Zealand was not permitted to export apples to Australia because of the presence of fire blight, a serious bacterial disease that was brought to New Zealand in the early twentieth century. Since the mid-1980s New Zealand has sought to have the fire blight ban lifted as there is no evidence that fruit carries the bacteria. After 20 years of failed negotiations, New Zealand decided to refer the matter to the World Trade Organization in 2007. Finally, in August 2011 the Australian government announced that New Zealand was permitted to supply apples provided that it followed a very strict biosecurity protocol to prevent potentially devastating pests and diseases, including fire blight, from entering the country.

Asia

There are certain market constraints and challenges around exporting pipfruit into Asia. These include import tariffs, phytosanitary issues (mainly relating to insects and disease), and a requirement for careful marketing to avoid oversupply. Asian markets have a preference for apples that are sweet and red-coloured, so there is demand for varieties such as Jazz™ and Braeburn.

References

Belrose. 2018. World Apple Review — 2018 Edition. www.e-belrose.com/apple-world-review.

Boniface, A. 2019. 'A is for Apple.' Westpac Over the Fence Economic Update. 7 June 2019. www.westpac.co.nz/assets/Business/Economic-Updates/2019/Monthly-Files-2019/NZ-Over-the-Fence-7-June-2019.pdf.

Elzebroek, A.T.G., Wind, K. 2008. *Guide to Cultivated Plants.* Wallingford: CAB International.

FAO and WHO. 2019. Codex 2019: The year of food safety. www.fao.org/3/ca5180en/ca5180en.pdf.

Ministry of Primary Industries. 2020. Situation and Outlook for Primary Industry (SOPI): December 2020. www.mpi.govt.nz/dmsdocument/43345-Situation-and-Outlook-for-Primary-Industries-SOPI-December-2020.

NZ Apples and Pears. 2019. www.applesandpears.nz/.

New Zealand Apples and Pears. 2020. Workforce Development Strategy. Personal communication.

Organic Exporters Association Research. 2018. 'Who We Are.' Accessed 19 February 2021. www.organictradenz.com/who-we-are.html.

Palmer, J. 2008. 'Apples and pears.' Accessed 18 February 2021. Te Ara — the Encyclopedia of New Zealand. www.TeAra.govt.nz/mi/apples-and-pears/print.

Plant and Food Research. 2013. 'New Apples Desired By Consumers.' Accessed 19 February 2021.

www.plantandfood.co.nz/growingfutures/case-studies/
new-apples-desired-by-consumers.

Plant and Food Research. 2017. Freshfacts — New
Zealand Horticulture. www.freshfacts.co.nz/files/
freshfacts-2017.pdf.

Plant and Food Research. 2018. Freshfacts — New
Zealand Horticulture. www.freshfacts.co.nz/files/
freshfacts-2018.pdf .

Plant and Food Research. 2019. Freshfacts — New
Zealand Horticulture. www.freshfacts.co.nz/files/
freshfacts-2019.pdf .

Statistics Portal. 2019. Agricultural Production Census:
Additional tables 2017. www.stats.govt.nz/information-
releases/agricultural-production-statistics-june-2017-
final.

T&G Global. 2021. ENZA. https://tandg.global/our-
produce/brands/enza.

USDA (United States Department of Agriculture). 2019.
Fresh apples, grapes, and pears: World markets and
trade. www.fas.usda.gov/data/fresh-apples-grapes-and-
pears-world-markets-and-trade.

Wiltshire, J.W. 2003. 'Integrated fruit production in the
New Zealand pipfruit industry. A report written for the
Primary Industry Council/Kellogg Rural Leadership
Programme, 2003.' http://researcharchive.lincoln.
ac.nz/bitstream/handle/10182/4385/Wiltshire_2003.
pdf?sequence=2&isAllowed=y.

Chapter 8

The Kiwifruit Industry

Svetla Sofkova-Bobcheva

Chapter 8
The Kiwifruit Industry

Svetla Sofkova-Bobcheva

School of Agriculture and Environment, Massey University

Introduction

Kiwifruit production is the largest horticultural industry in New Zealand. It generates the highest percentage of export revenue of all the horticultural industries, supports hundreds of businesses and employs thousands of people along its export and domestic value chains. New Zealand is the pioneer of this industry and globally is one of the top three producers. The climate and physical conditions across many parts of New Zealand are well suited to the cultivation of kiwifruit, and there is much scope for innovation to continue to grow this industry and to remain competitive with other countries.

On a global scale, kiwifruit is a niche fruit taking up an estimated 0.22 per cent of the global fruit bowl. Apples, oranges and bananas dominate these markets. In New Zealand, however, kiwifruit is the single largest horticultural export product, recently earning more than $2.3 billion annually. Export revenue reached $2.2 billion for the year ending March 2019, an increase of 36 per cent over the previous year. The increase was due to excellent weather in 2018 which increased yields, maturation of the new Gold variety plantings, and prices for Green and Gold varieties increasing by, on average, 11 per cent on the previous year (Ministry of Primary Industries 2019). The kiwifruit marketer Zespri expects to achieve more than $3 billion in sales in 2020–21 and is on track to reach $4.5 billion in five years, supported by increasing global volumes of kiwifruit.

In 2017, New Zealand was ranked third in terms of kiwifruit production, producing 411.78 thousand metric tons (MT), just behind Italy at 541.15 thousand MT and China at 2024.6 thousand MT (Statista 2019).

The current state of the kiwifruit industry in New Zealand

The kiwifruit industry is currently the most important horticultural industry in New Zealand, representing around 33 per cent of total horticultural export revenue. Kiwifruit production has increased by 50 per cent over the past decade (Figure 8.1) and forecast revenue is NZ$4 billion by 2027. New Zealand exports to 50 countries, with the largest markets being China, Japan, Spain, Taiwan, Germany and South Korea (Figure 8.2). Of these, in 2019 revenue of $1.53 billion came from Asian markets (Japan $590 million, China $510 million). The New Zealand government and the kiwifruit industry jointly invest $20 million per year in the new varieties breeding program (NZKGI 2018).

In the 2019/20 season New Zealand had 12,906 hectares (ha) planted in kiwifruit, with 10,743 ha in the Bay of Plenty (Table 8.1), and there were about 2600 growers and 2900 registered orchards in New Zealand. In the 1960s and 1970s the kiwifruit industry went through a boom period and much land was planted in kiwifruit. However, since then the industry experienced some hard times due to bacterial attack and there were significant declines in kiwifruit production in the Hawke's Bay, Gisborne and Nelson regions. Waikato and Auckland are now the second and third largest producers of kiwifruit in New Zealand (Table 8.1).

Glossary

Hayward first cultivar produced by Hayward Wright in New Zealand in 1927

dioecious a plant having the male and female reproductive organs in separate individuals

NZKA New Zealand Kiwifruit Authority

SPE single point of entry

KVH Kiwifruit Vine Health

NZKGI New Zealand Kiwifruit Growers Incorporated

KNZ Kiwifruit New Zealand

IAG Industry Advisory Council

ISG Industry Supply Group

Psa *Pseudomonas syringae* pv. *actinidiae* (canker disease)

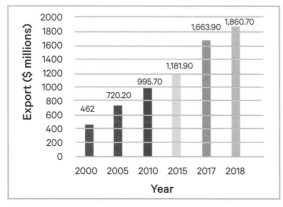

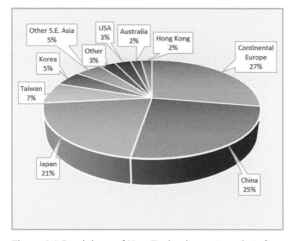

Figure 8.1 Orchard-gate return and growth in export value for the kiwifruit industry. (Source: Plant and Food Research 2018)

Figure 8.2 Breakdown of New Zealand export markets for the year to June 2018. (Source: Fresh Facts 2019)

Origin and development of the kiwifruit industry

Domestication and use of kiwifruit

New Zealand can take credit for pioneering the kiwifruit industry, having created the first cultivar from wild Chinese origins and developing many of the horticultural techniques used in commercial horticulture today.

Kiwifruit are native to southwest China where they have the common names yang tao, mihoutao or monkey peach, and Chinese gooseberry. In this region temperatures are warm and winters are mild, and kiwifruit vines climb into forest canopies and along the ground and form dense stands. The Chinese have collected the fruit for thousands of years but never domesticated the plant (Ferguson and Huang 2007).

Kiwifruit belong to the genus *Actinidia* and the family Actinidiaceae. The wild variety is known by the scientific name *Actinidia chinensis* and has small, smooth-skinned fruit. Cultivated kiwifruit have the scientific name *Actinidia deliciosa* and have larger and hairier fruit. All species in the genus *Actinidia* have common morphological features, including a climbing growth habit, dioecy (having male pollinating and female fruit plants), a characteristic radial arrangement of styles, and similar structure of fruit (Ferguson 2013).

Botanists who visited China in the early 1900s introduced seed and plants of both species (*A. chinensis* and *A. deliciosa*) into Europe, the United States and New Zealand, but only *A. deliciosa* established successfully. In Europe and the United States this did not result in a

Table 8.1 Regional production of kiwifruit in the 2019 season.

Region	Area (ha)
Bay of Plenty	10,743
Waikato	552
Auckland	495
Northland	477
South Island	73
Poverty Bay	354
Hawke's Bay	212
Total	**12,906**

(Source: Zespri 2020)

commercial crop, being grown by enthusiastic gardeners only; possibly because plants of both sexes were required. Seeds taken to Australia in 1904 must have produced both male and female plants, as fruit was being produced there by 1910 (Ferguson and Bollard 1990).

In China, kiwifruit (both *A. chinensis* and *A. deliciosa*) is collected from the wild in large quantities — possibly equal to that cultivated globally. It is sold locally for fresh consumption and also processed into a range of products such as fruit juices, jams, preserved fruit, wine and spirits; many other parts of the plant are used for a variety of purposes. The stems make good-quality paper and adhesives and the leaves are fed to livestock. Fragrant oils are extracted from the flowers and seeds. There are also many traditional Chinese pharmaceutical uses of these species.

Globally, the primary use of the commercially cultivated crop is fruit for fresh consumption. The fruit has an appealing bright green colour, a unique flavour and a very high vitamin C content. In New Zealand, about 75–80 per cent of the fruit produced meets export standards. As a result, as production has increased, so too has the volume of fruit available for processing. However, in processing the hairy skin is hard to remove and the qualities of the fruit that appeal to the fresh food market are hard to preserve. The chlorophyll that gives the fruit its appealing green colour degrades to a less appealing olive green, the turgidity is lost and the pulp becomes more of a sludge, and both flavour and vitamin C are lost with processing and heating. Despite these processing issues, New Zealand exports a variety of processed products including canned sliced fruit, frozen fruit (slices or pulp), wine, jam, juice, juice concentrate, glazed fruit and other products.

Cultivation in New Zealand

Kiwifruit was first brought to New Zealand in 1904 as seed from China by a teacher from Whanganui called Isabel Fraser. A horticulturalist called Alexander Allison propagated this seed and started growing kiwifruit.

In 1927, a cultivar was bred by Hayward Wright in the Bay of Plenty. This variety was named the 'Hayward'. The name 'kiwifruit' was first used in 1959 by the New Zealand fruit-handling firm Turners and Growers when they first exported *A. deliciosa* to the United States (Ferguson and Bollard 1990). By the 1960s the 'Hayward' was the main variety in New Zealand, and by 1975 it was the only fruit recommended for export (Ferguson 2013). This variety makes up 90 per cent of world kiwifruit production. The large fruit size, excellent flavour and storage qualities are the main attributes that have contributed to this variety dominating the world market and positioning New Zealand as one of the leading suppliers in the international kiwifruit market.

The New Zealand kiwifruit industry is essentially a monoculture because it is based on one pistillate (female-bearing) cultivar. Not only does this carry risk because of the possible impact of disease, but it also creates logistical problems as the harvest period occurs over just 6 weeks. Over time, various strains of Hayward have been developed, many of which are grown around the world. The Chico cultivar was sent to California in 1935 by Hayward Wright and became the source of much of the commercial plantings there. Other varieties from New Zealand are now grown in France, Italy and Japan. Growers in California have managed to achieve good yields with cultivars such as Abbott, Bruno and Monty, although these comprise <1 per cent of world production.

Structure and governance

There have been several waves of growth in New Zealand's kiwifruit industry since the first cultivar was developed in the early 1900s. Linked to these have been the formation of industry structures and processes that have managed this growth and maintained quality and competitiveness in international markets.

The commercialisation of kiwifruit took place in the 1960s, leading to the Kiwifruit Export Promotion Committee being established in the 1970s, which in turn resulted in the establishment of the New Zealand Kiwifruit Authority (NZKA) in 1977. The main role of the NZKA was to coordinate packaging, maintain an export grade and manage promotions. They also had to manage export licenses as there were up to seven exporters licensed to operate.

In the 1980s a few events had great impacts on the kiwifruit industry. In the early to mid-1980s there was a boom in production, and export sales rose from 22,000 tonnes in 1981 to 203,000 tonnes by 1987. However, this increase in volume led to a drop in the price per tray from $7.84 in 1981 to $3 in 1987, with 91 per cent of the growers making a loss from their kiwifruit operations at the end of that period. In 1987, the New Zealand dollar also rose sharply in value and interest rates rose to 20.5 per cent. There was a big drop in international demand for New Zealand kiwifruit because it had become so expensive due to the high New Zealand dollar.

The kiwifruit industry was now in crisis, and as a result land values dropped along with returns. This chain of events started a debate about the advantages of having a single exporter instead of multiple exporters so that price and quality could be better moderated. In September 1988, a referendum was held among growers to gauge support for a single point of entry (SPE) because of the falling export prices and undercutting between the seven kiwifruit exporters. In 1989/90, 84 per cent of growers supported the formation of the Kiwifruit Marketing Board, which had statutory power to buy all kiwifruit that was to be exported (NZKGI 2018).

The mid-1990s saw a review of the kiwifruit industry in New Zealand, which recommended major structural change. The first stage of this change involved the formation of New Zealand Kiwifruit Growers Incorporated (NZKGI), which became operational from 1994. The second stage involved major changes in marketing and branding, which led to the formation of the Zespri brand. This was launched in 1996/97, with Zespri being set up as a separate marketing and sales organisation. The third and final stage involved a review of corporatisation, collaborative

marketing and industry operational structure. This led to another grower referendum, and thereafter changes resulted in Zespri becoming the marketing company of NZKGI and NZKMB. All of these organisations are still present today. The review also recommended the growth of new varieties and the development of plant breeding, and produced Zespri's strategy to become the world's leading marketer of kiwifruit throughout the year.

Zespri International Limited

Zespri was launched and officially corporatised in 2000. All growers at that time became shareholders, with shares being allocated according to the number of trays produced by each grower. In 2001 a change to kiwifruit legislation saw the introduction of a voting cap to ensure that the growers retained control of the industry (NZKGI 2018).

In a referendum in 2015, 98 per cent of the growers voted to keep the SPE structure. This structure allows for one exporter only. Zespri holds the SPE for the kiwifruit industry. The marketing structure helps producers deliver scale in the marketplace and allows Zespri to choose the distributors to serve each market. There are many other benefits of this marketing structure, such as ongoing investment, branding for premium product, commercialisation of new varieties, controls for consistent quality, good customer service, sustainability and competitive returns (NZKGI 2019).

Zespri International Ltd is now the world's largest marketer of kiwifruit, selling into more than 50 countries and managing 30 per cent of the global volume of trade in kiwifruit. It has set the global benchmark for guaranteed excellence, and the Zespri brand is a guarantee of top-quality, delicious kiwifruit. Zespri is responsible for marketing approximately 98 per cent of New Zealand's export crop of kiwifruit, other than the fruit sold to Australia which is handled by collaborative marketers. In 2015 it was estimated that approximately $4000 million was invested in orchards, $310 million in post-harvest firms and $460 million in Zespri (Scrimgeour-Locke 2015); this value will have grown substantially since then.

The Zespri brand was launched in 1996/97, and in 1997 Zespri Gold, the first alternative variety to the Hayward cultivar, was launched. Currently, Plant and Food Research is responsible for managing the biggest kiwifruit breeding programme and germplasm collection in the world. Zespri has long been the largest client of Plant and Food Research, investing several million annually into research and development, which includes a kiwifruit cultivar development programme, health research, postharvest science, consumer and sensory testing and sustainability research. When Psa (*Pseudomonas syringae* pv. *actinidiae*) — the causal agent of bacterial canker disease — was found in New Zealand in November 2010, more than 100 Plant and Food Research staff were redeployed to the fight against this devastating disease. In the six months following the discovery of Psa, Plant and Food Research conducted more research about the pathogen than had ever been done before. This rapid research helped develop management strategies, detection methods and molecular diagnosis that played an important role in supporting Zespri's management of the pathogen and gave confidence to international marketplaces not to restrict market access. The joint cultivar programme identified a Psa-tolerant Gold cultivar: Gold3, now marketed as

SunGold, proved to be the basis of recovery for the New Zealand kiwifruit industry.

More recently, other new varieties have been introduced (Figure 8.3). These include a new green-flesh kiwifruit (Sweet Green) and a red-flesh variety (Zespri Red), which were developed along with other innovations that allow earlier harvesting and longer storage to extend the potential selling period. The advantages of the new green-flesh and red-flesh varieties include the fact that they are sweeter than the older varieties and therefore appeal to a different palate, especially the Asian market.

The overarching goal of Zespri is to fill the gaps in the supply of kiwifruit from New Zealand to international markets and to provide consumers with Zespri-branded kiwifruit all year round. Zespri is expanding production outside of New Zealand to meet demand for the SunGold variety, as a key part of the marketer's growth strategy is expanding its 12-month supply programme. To achieve this, Zespri has partnered with growers in the northern hemisphere. Zespri has licensed growers in Italy, France, Japan and South Korea to grow Zespri Gold and SunGold kiwifruit.

Zespri's SunGold variety has been a consistently strong performer in the market over the past few seasons. The marketer has speculated that future demand for kiwifruit will expand, and an additional 3500 ha of SunGold licences are being released from 2018 onwards: each year until 2022, an additional 400 ha of SunGold and 50 ha of SunGold organics will be released for tender. This will add to the 5000 ha of SunGold currently grown across the country (Plant and Food Research 2018).

Zespri handles around one-third of the global trade in kiwifruit and has built a strong reputation by focusing on marketing, developing new

Figure 8.3 Some of the main Zespri kiwifruit varieties: Zespri Red, SunGold and Green.

products, innovation, brand awareness, taste and consistency, and supply chain development.

Other organisations

Other key organisations involved in the kiwifruit industry form the basis of how the industry operates today:

- Kiwifruit Vine Health (KVH) is funded through a national pest management levy for the management of Psa (0.6 cents per tray). The levy is used to fund biosecurity readiness and response.
- Kiwifruit New Zealand (KNZ) was established under the Kiwifruit Export Regulations 1999 (New Zealand Government, 2014) and is the regulator of the New Zealand kiwifruit industry.
- New Zealand Kiwifruit Growers Incorporated (NZKGI) was formed following the downturn in the industry in 1993 to give growers their own organisation and voice to develop a secure and stable kiwifruit industry.
- Māori Kiwifruit Growers Forum was established in 2016 by Māori for Māori to build and grow the Māori kiwifruit business for current and future generations.
- Kiwifruit Vine Health (KVH) is a biosecurity organisation established in 2010 to lead the response to the Psa incursion. Since 2012, KVH has been the organisation responsible for managing all biosecurity readiness, response, and operations on behalf of the kiwifruit industry, working collaboratively with growers, Zespri, NZKGI, the postharvest and associated industries, and Government.
- The Industry Advisory Council (IAC) aims to specifically cater to the financial, tax and Government-related aspects of the kiwifruit industry. IAC manages issues relating to the

Supply Contract, decisions relating to the treatment of and payment for fruit, and matters with material financial implications for growers.

- The Industry Supply Group (ISG) manages decisions relating to the supply chain process, monitoring quality assurance, labelling, packaging and the export of kiwifruit. ISG also helps in the negotiation of industry-wide commercial contracts relating to supply chain activities.

On-orchard practices

Establishment

Site selection

Kiwifruit requires fairly specific growing conditions. It requires warm, sunny summers to accumulate dry matter, but it also needs sufficient winter chill — below 7°C — for good floral development and bud burst. The best orchard sites are north facing, with those facing east or west being the next preference. South-facing orchards have less sunlight and greater exposure to cold winds. At high elevations, conditions become colder and wetter; 200 m above sea-level is the upper limit for orchards. Orchard topography must not be too steep, as this makes the use of machinery perilous.

Climate and soil

New Zealand has excellent conditions for growing kiwifruit, which contribute to its unique quality and flavour. They include clean air, fertile soils, a suitable climate and few pests and diseases. The main factors required for growing kiwifruit include plentiful sunshine, water, rich free-draining soils (pH 5–6.8) and winter chilling. These are particularly available in the North Island. Kiwifruit do not like wet feet, so soils need to be well drained. Water-holding capacity, however, is important and this is achieved in soils having high levels of organic matter.

Access to a source of water for irrigation is important for the development of young plants and during dry periods. Vines lose a lot of water during the height of summer and irrigation is essential. Water may also be required for frost protection during winter. There are often regional and local council restrictions on water use, so having access to a source of water or an ability to store it is important.

Establishment costs

Important economic considerations regarding establishment costs include the initial capital cost required to set up the orchard and the time taken before income is generated. The time from planting an orchard to fruit production may be 3–5 years with no returns before the orchard starts fruiting and covering annual growing costs. Depending on the site, the total cost of orchard establishment may vary between $120,000 and $400,000 per hectare.

In addition, the median price of a kiwifruit licences in 2019 was $290,000 per hectare; some prime kiwifruit orchard country sells for $1.3 million per hectare (Ministry of Primary Industries 2019). Thus, considerable investment is required to establish new gold kiwifruit orchards. In some areas where land prices are extremely high, such as the Bay of Plenty, some orchardists may look to higher-value crops such as berryfruit or avocadoes (Ministry of Primary Industries 2019).

Establishment costs include:

- **Site preparation.** This might involve contouring the area to make management easier.
- **The establishment of shelter before vines are planted.** Shelter is required to reduce the impacts of wind, which include damage wind rub and evapotranspiration. Moreover, cold temperatures can reduce flowering and bee activity. Tree shelterbelts take time to establish, while artificial windbreaks can increase shelter but do not limit light. The latter are more expensive but more immediate. Overhead shelters reduce the impact of hail, reduce wind speed and turbulence, reduce leaf wetness and increase pest control.
- **The establishment of water supply and reticulation.** Young vines require constant watering to help leaf growth and root system development. The soil type will determine how much water is required. Soils high in pumice require more water than do clay soils.
- **The planting of rootstock and a year later grafting of a variety.** A female kiwifruit variety (the scion) is grafted on to another type of female kiwifruit, the rootstock.
- **Building support structures.** Vines have to be trained on to support structures. The most common system is the pergola system. Stringing increases the growth of new leaders (some continue every season and others choose a low-vigour system).

Frost protection using water or windmills and overhead hail protection may need to be established. All significant frosts in New Zealand are the radiation type, which occur at night with clear skies and little wind. Heat is radiated away from the surface of vegetation, and the surface cools and draws heat from the plant material and surrounding air until the temperature of the plant material falls and causes irreversible damage. Frost-damaged fruit is not edible or saleable.

Frost-protection systems include heating using 'frost pots', sprinklers, wind machines on a 10-metre tower, flying a helicopter low over vines, radiation barriers (frost cloth, fog) and cold-air drainage (cold air is denser and heavier).

Another important consideration is the location of the plantation. In areas where a good service industry already exists, there is more opportunity to use contract services; but if the orchard is located away from such services then greater infrastructure is needed to house equipment and labour.

Orchard management practices

The kiwifruit is a vigorous, long-lived, perennial deciduous vine, and in its mature form it requires a strong framework on which to grow. The plants can live for over 50 years. Under most systems of orchard management, the vine is grown to about 1.8 m in height and can be more than 20 cm in diameter.

A mature orchard kiwifruit vine has a permanent framework of branches, and from these fruiting canes develop at right-angles (Figure 8.4). The vine arms with current-year lateral shoots and second-order lateral shoots are the typical productive units of kiwifruit vines.

Kiwifruit have an extensive root system, and the fresh weight of the roots can approach that of the total weight of the vine above ground. In deep, porous soils the roots can grow in large numbers to a depth of at least 4 m; in coarse sandy soils the roots can go down to more than 2–4 m. They extract significant amounts of water. Roots usually extend 2–3 m from the trunk,

Figure 8.4 Kiwifruit orchard: vines growing on a T-bar trellis.

and when established overlap with the root systems of adjacent vines.

A number of management practices are carried out throughout the year (Table 8.2). Some relate to optimising sunlight and airflow to maximise growth and reduce disease through vine training and pruning, while others are carried out to maximise yield through good pollination. Work is also carried out to maintain the structure of the orchard to provide shelter from winds, pests and hail.

- *Thinning* is done to reduce the number of fruit so as to increase fruit size and tastiness. This is done in summer.
- *Girdling* — trunk girding is done to increase dry matter, increase fruit weight and increase the number of flowers the following season. This is done in spring and summer.
- *Root pruning* is done to reduce the size of the root systems and reduce the carbohydrate demand of the root system. This is done to increase fruit growth and dry-matter accumulation.
- *Orchard risk management practices* — these include all of the factors that can affect the profitability and sustainability of the orchard. They include the consideration of increasing growing costs, downturns in returns, increase in labour, natural events, etc. A number of factors that relate to crop protection also require consideration, including integrated pest control, agrichemical control and orchard hygiene.

Flower structure, composition and pollination

The flowers of kiwifruit are dioecious, which means that individual plants are either male or female. In fact, the flowers of pistillate (female) kiwifruit have fully functional male parts (stamens) and female parts (pistols), but the stamens (male parts) do not produce viable pollen. The staminate plants have a greatly reduced ovary and only poorly developed styles. It is the physical structure of these flowers that limits self-fertilisation. Male plants occasionally produce fruit, and experiments have demonstrated that such plants have flowers that are self-pollinating and self-setting. This means that there is potential for a truly hermaphroditic (both sexes) cultivar to be produced.

Pollination is a very important aspect of kiwifruit management, as the yield is dependent on the number and size of the fruit. There are major differences between varieties in pollination requirements. Hayward flowers need up to 40 bee visits for full pollination, while Gold3 (SunGold) only requires six (Zespri 2016). Most of the male and female flowers open before 8 a.m. The pollen starts to be released from 8–9 a.m. and is strongly released until about 12 p.m., when it drops off, generally stopping by evening. The optimum temperature for pollen grain development is between 12 and 26°C.

A number of factors affect the pollination of kiwifruit flowers. Pollination of kiwifruit requires greater management than most other crops, and this increases costs depending on the pollination technique used. Kiwifruit have relatively few flowers and require >80 per cent fruit set (higher than pip and stone fruits). The flowers are not attractive to insect pollinators because they do not produce nectar, and only 6 per cent of bees placed in orchards will visit kiwifruit flowers (Zespri 2017). Because kiwifruit originated in China, it naturally did not evolve to use New Zealand insect pollinators. Pollen also has to be moved relatively long distances, as

Table 8.2 Kiwifruit orchard management practices.

Winter (June–August)

Vine pruning takes place while vines are dormant; in winter at end of previous year's harvest.

The aim of winter pruning is to set up bays with optimal high-quality winter bud numbers and canes that are evenly spaced.

Spring (September–November)

Vines grow, produce new shoots and first flower buds. Flowers blossom, bees pollinate and fruit is produced.

Work includes:
- Selective pruning of fruit shoots

- Tip squeezing to damage actively growing shoots to retard primary growth and stimulate growth from lateral buds

Summer (December–February)

In summer the kiwifruit vines show tremendous growth, and pruning is required to direct growth and manage the canopy. Pruning removes excess vegetation growth to allow light in.

Fruit are selectively thinned to optimise size and taste (less fruit, increased size and tastiness).

Autumn (March–May)

As weather cools, harvest approaches.

Kiwifruit is tested for ripeness, and when they pass specific criteria for quality and grade they are carefully picked.

Fruit is transported to packhouses and packed and stored for shipping and export.

As weather gets colder, leaves drop and vines move towards a dormant stage.

male and female flowers are found on different vines. Fruit size is determined by the number of seeds that the fruit contains, which is dependent on the level of pollination. Each female flower needs to receive thousands of pollen grains.

The vines are generally grown in wind shelters, which reduces wind pollination. Growers supplement bee and wind pollination with artificial pollination. There are three different techniques: contact application (hand pollination), wet application (spraying pollen on to flowers), and dry pollination (blowing pollen).

Harvest, postharvest and marketing

Fruit can only be harvested when it meets a minimum dry-matter threshold, and has the right colour, at least the minimum soluble solid concentration (SSC) or brix (sugar content), and sufficient black seeds. Dry matter is the most important aspect of fruit maturity, and fruit payments will depend largely on this (Taste Zespri grade) and SSC.

Harvesting starts in late March, peaks in May and is over in early June. Zespri extends the season by paying a premium, known as Kiwistart Premium, for fruit picked early — these fruit are generally smaller and may not have the full taste. There are also payments for fruit sold later to compensate for extra freezer time. Fruit can be stored for months by optimising inventory management practice, fruit maturity and high-quality fruit handling.

There are 46 packing facilities and more than 95 cool-stores used by the kiwifruit industry across New Zealand. Packing is a key control point where fruit is sorted into market-acceptable product. At this point, Global Good agreement practices and market restrictions come together. Fruit is labelled, placed in cool storage and shipping is organised.

Future challenges and opportunities

Technology

In pre-harvest operations, drones, GPS units and GIS software are increasingly being used to monitor crop condition. Ground-based digital crop-counting techniques (to count flower buds, flowers and fruit) are also making use of new technology.

A major challenge relating to pollination is that overhead nets impede the sun's rays and affect the bee's ability to operate. Zespri and Plant and Food Research are trialling different options for pollination (Zespri 2017).

Autonomous machines are now available for harvesting (Figure 8.5).

Post-harvest there are many ways in which the industry is using technology. These include:
- Fruit quality control, labelling and sorting is becoming more and more automated.
- Infrared cameras are being used for grading.
- Robots are used for packing and stacking.
- Automated temperature monitoring and management of fruit is becoming part of the movement of fruit from cool-store to market (Figure 8.6).

These new technologies are useful given that there is competition for labour, and they mark a shift to more-efficient, higher-quality practices.

Labour requirements

Skilled labour for pruning and picking of kiwifruit is very important, and the industry is reliant on imported labour from the Pacific Islands as part of the Recognised Seasonal Employment (RSE) scheme. In 2017, NZKGI reported 15,678 seasonal employees in the kiwifruit industry; of these, 10,000 people were employed all year. This number was predicted to grow to 22,699 by 2027. The RSE scheme increased from 5000 overseas workers in 2007 to 11,100 in December 2017.

In 2020 the Covid-19 pandemic severely impacted the country's seasonal workforce, with very limited RSE workers and backpackers able to enter New Zealand. Pacific workers who would normally travel to New Zealand were tempted to Australia for their recommended Seasonal Worker Programme. Some of the suggestions to cope with the labour crisis include:

- Attracting unemployed New Zealanders, school leavers, and tertiary students into the horticulture sector for seasonal work
- Utilising prisoners from corrections facilities who are eligible for day release
- Offering increased flexibility of work hours
- Adapting the workplace using technologies such as work platforms to lower the physical requirements of the work.

Biosecurity

Pseudomonas syringae pv. *actinidiae* (Psa) causes bacterial canker of kiwifruit, and in 2017 was the most threatening disease of the Actinidiae species worldwide (Tyson et al. 2018). This disease had a significant impact on the industry in 2013/14. The Psa canker disease is specific to kiwifruit and in 2014 was responsible for a major decline in production. There was a significant reduction in Gold kiwifruit production from 2012 to 2014, and significant industry stress during this period. The Gold3 (SunGold) variety of kiwifruit, which replaced the Hort16A variety, is resistant to Psa, and along with improved management practice the impact of Psa was reduced; the industry has grown strongly ever

Figure 8.5 Autonomous vehicles that can navigate in orchards can monitor and harvest crops around the clock.

Figure 8.6 Zespri SunGold kiwifruit quality control in packhouse and cool-store.

since. It continues to expand and the planted area has grown to over 13,250 ha currently.

In 2018 kiwifruit growers were able to access a new tool to help them combat Psa. This is a biocontrol product called AureoGold™, a yeast found naturally on many different species of plants and fruits in New Zealand that is safe for bees. This natural yeast strain reduces the growth and spread of Psa and was developed by Plant and Food Research in collaboration with Arysta LifeScience. It is used during flowering and following fruit set, which is a time in the calendar when use of other controls for Psa is limited (Plant and Food Research 2018).

Brown marmorated stink bug, an insect in the family Pentatomidae, is native to China, Japan, the Korean peninsula and Taiwan. It is an agricultural pest that is found mainly in Asia, Europe and the United States. While it has not yet been found in New Zealand, it would have a massive impact on horticulture as it feeds on fruit trees, field crops and woody ornamentals. It would affect a number of New Zealand's horticulture sectors, including seeds, vegetables, kiwifruit, apples and pears, and wine.

The introduction of this pest into New Zealand would entail a significant cost to the country's economy from yield losses, pesticide costs, reduced on-orchard productivity due to managing and monitoring the pest, possible reduced export prices due to consumer perceptions about quality, and the growth of sacrificial crops to protect those that are more valuable (NZIER 2017). There is a biological control agent, the samurai wasp, that could be introduced with Environmental Protection Authority approval. Estimates of the costs to the industry and management actions have been conducted (NZIER 2017).

Environmental risks

In 2010 Zespri developed a strategy to manage environmental risks to the kiwifruit industry. The top five risks identified were:

- greenhouse gas emissions (carbon footprint)
- water
- waste
- non-renewable resources
- biodiversity.

The kiwifruit industry has relatively low use of nitrogen compared with other land uses, but is still trying to model nutrient loss and impact to develop best practices.

In the future, it is likely that climate change will have an impact on the kiwifruit industry. Kiwifruit dormancy and flower numbers require a sufficient winter chilling period, and as global average temperatures rise there is a risk of insufficient winter chilling. A study conducted by Tait et al. (2017) on the Hayward cultivar in Te Puke, Bay of Plenty, shows that production is set to decrease over the coming decades, and could become non-viable by the end of the century under all but the most stringent of the global greenhouse gas emissions pathways. There are, however, many other areas across New Zealand that show a potential increase in Hayward kiwifruit production (Tait et al. 2018).

References

Ferguson, A.R. 2013. Kiwifruit: The wild and cultivated plants. *Advances in Food and Nutrition Research* 68: 15–32. doi: 10.1016/B978-0-12-394294-4.00002-X

Ferguson, A.R., and Bollard, E.G. 1990. Domestication of the kiwifruit. In I. J. Warrington and G. C. Weston (eds), *Kiwifruit: science and management*. Auckland: Ray Richards Publisher and New Zealand Society for Horticultural Science, pp. 165–246.

Ferguson, A.R., and Huang, H. 2007. 'Genetic resources of kiwifruit: Domestication and breeding.' *Horticultural Reviews* 33: 1–121. doi: 10.1002/9780470168011.ch1

Ministry of Primary Industries. 2019. 'Situation and Outlook for Primary Industries (SOPI), March 2019.' Accessed 202 December 2020. www.mpi.govt.nz/resources-and-forms/economic-intelligence/situation-and-outlook-for-primary-industries/sopi-reports/.

NZIER. 2017. 'Quantifying the economic impacts of a Brown Marmorated Stink Bug incursion in New Zealand. A dynamic Computable General Equilibrium modelling assessment.' Accessed 25 September 2019. www.hortnz.co.nz/assets/UploadsNew/Quantifying-the-economic-impacts-of-a-Brown-Marmorated-Stink-Bug-Incursion.pdf.

NZKGI. 2018. 2018 *Kiwifruit Book*. Accessed 22 December 2020. nzkgi.org.nz/wp-content/uploads/2016/12/2017-kiwifruit-book-22-10.compressed-1.pdf.

Plant and Food Research. *2018* (29 November). 'New biological control for kiwifruit.' Accessed 22 December 2020. www.plantandfood.co.nz/page/news/media-release/story/new-biological-control-for-kiwifruit-disease/.

Statista. 2019. www.statista.com.

Tait, A., Paul, V., Sood, A., and Mowat, A.D. 2017. 'Potential impact of climate change on Hayward kiwifruit production variability in New Zealand.' *New Zealand Journal of Crop and Horticultural Science* 46(3): 175–197. doi: 10.1080/01140671.2017.1368672

Tyson, J.L., Vergara, M.J., Butler, R.C., Seelye, J.F., and Morgan, E.R. 2018. 'Survival, growth and detection of *Pseudomonas syringae* pv. *actinidiae* in *Actinidia in vitro* cultures.' *New Zealand Journal of Crop and Horticultural Science* 46(4): 319–333. doi: 10.1080/01140671.2017.1414064

Zespri. 2016. Annual Review. https://assets.ctfassets.net/b7rvvweqeqmn/68djLsAeukyvnxDng4TWGI/d18320f-2cfe64418815108bd8d6503b2/Annual-Review-2015-16.pdf.

Zespri. 2017. Annual Report. https://assets.ctfassets.net/b7rvvweqeqmn/1HaBvvTrbo0T6LWYk5HwmX/10c25f-1c46ff94f05b0cb86d8f4a79ff/Annual-Report-2016-17.pdf.

Zespri. 2020. Annual Report. https://assets.ctfassets.net/b7rvvweqeqmn/4BPJQeCk9y30NvmWT8yheO/ca04475a4511b08e6ce4bdce57a506d3/Annual-Report-2019-20.pdf.

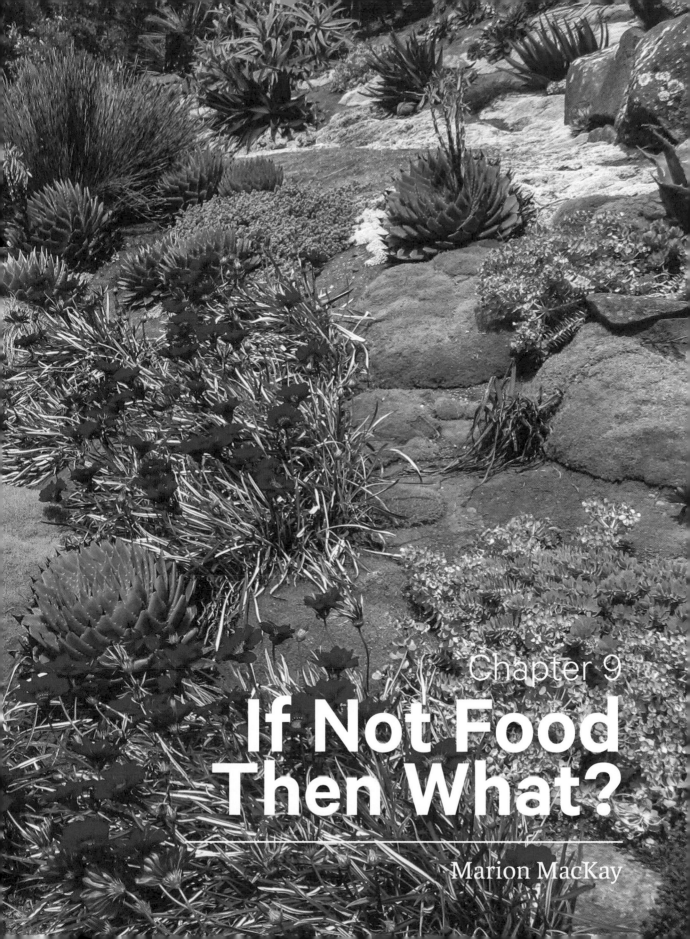

Chapter 9

If Not Food
Then What?

Marion MacKay

Chapter 9

If Not Food Then What?

Marion MacKay

School of Agriculture and Environment, Massey University

Introduction

Horticulture has a range of applications other than food production, including conservation, urban horticulture, parks and gardens, and production of cut flowers and foliage. These are underpinned by plant propagation (including selection and breeding) and growing of plants by the nursery industry. As for food-production horticulture, aspects such as growing substrate, soil fertility, climate, pruning, flowering and seeding are important, but often in different ways.

Conservation

Conservation is an important issue. There is a global problem with loss of plant biodiversity — internationally, about a quarter of plants are threatened with extinction (Royal Botanic Garden Kew [RBG Kew] n.d. [b]; Botanic Gardens Conservation International [BGCI] n.d. [a]), while in New Zealand 31 per cent of plant species are classified as 'At Risk' and another 15 per cent are 'Threatened' (de Lange et al. 2018). Although plants should ideally be conserved within their native habitat (in situ conservation), often this is simply not possible because the native habitat has been destroyed or converted to some other form (e.g. agricultural or industrial uses) and is no longer available. Horticultural cultivation is used to conserve plants ex situ (out of habitat), usually in living collections in botanic gardens, or in seed banks.

Botanic gardens are centres for the cultivation of plants and research into that cultivation. Their original purpose was to 'discover' and figure out how to grow 'new' plants, including food plants, and identify economic uses of those species. Many plants that are now used for food or medicines were introduced to cultivation via a botanic garden. Today botanic gardens tend to focus on conservation, with international programmes in conservation led by Botanic Gardens Conservation International (BGCI) in London (see the end of the chapter for website details). Under the umbrella of the International Convention on Biological Diversity (BGCI n.d. [b]), BGCI leads two of the Targets of the Global Strategy for Plant Conservation (BGCI n.d. [c]): the conservation assessments of Target 2 and the ex situ conservation programmes of Target 8. New Zealand people and gardens are involved in both of these programmes.

New Zealand has botanic gardens in Auckland, Wellington, Christchurch and Dunedin, with smaller gardens in several other centres. These gardens contain a range of native and exotic plants, mostly grown outdoors but also with some glasshouse cultivation. Typical features include areas of native flora, rose gardens, displays of flowering annuals, collections of trees and shrubs, and particular collections of different kinds. For example, Dunedin Botanic Garden has an extensive collection of *Rhododendron* while Wellington Botanic Garden has a conifer collection. Indeed part of the conifer collection at Wellington was an early experimental planting to determine how *Pinus radiata* would grow in New Zealand. Because our botanic gardens are funded by local authorities, and must also

Botanic gardens

Don't be fooled by the term 'garden' when talking about botanic gardens. While all botanic gardens have an aesthetic display function and members of the public visit the garden to enjoy the plant displays, the key purpose of most botanic gardens today is scientific horticulture and conservation.

Two of the largest botanic gardens globally are the Royal Botanic Gardens Edinburgh and Kew in Britain, which are significant centres for science. Both grow thousands of different species of plants, have extensive ex situ conservation programmes, collaborate on in situ conservation programmes, undertake research programmes and publish scientific journals.

function as local parks, they tend not to have extensive science functions; however, staff at all of the gardens are involved in conservation projects of some form.

One garden, Otari Native Botanic Garden in Wellington, is the only botanic garden in New Zealand to grow just native species. As well as displaying common species and species suited to general garden cultivation, Otari is active in ex situ conservation of New Zealand flora and grows and collects rare species. Some of these are held in the collections of living plants, where one of the key horticultural challenges is figuring out how to grow species that often come from very particular habitats and can be tricky to grow in cultivation.

Even though some species are hard to grow, cultivation in gardens has 'saved' several New

If you want to grow it you have to know it!

Celmisia spedenii, Speden's mountain daisy. Conservation Assessment: At Risk — Naturally Uncommon. Seen here in cultivation at Otari Native Botanic Garden, Wellington.

Ranunculus paucifolius, Castle Hill buttercup. Conservation Assessment: Threatened — Nationally Critical. Seen here in habitat at Castle Hill.

Pennantia baylisiana, Three King's kaikōmako. Conservation Assessment: Threatened — Nationally Critical. Seen here at Massey University, Palmerston North.

Ultramafic only

This alpine daisy is only found on ultramafic ground — an unusual substrate that is high in magnesium and hosts particular ranges of plant species. Even more unusually, this daisy might be an obligate ultramafic (Lee & Reeves 1989): that is, it *needs* regular magnesium to thrive (New Zealand Plant Conservation Network [NZPCN], n.d. [a]). It grows poorly in typical garden soil, but grows well and produces seedlings when grown in soil from its habitat (Lee & Reeves 1989). *C. spedenii* comes from West Dome and the Livingstone range in Southland, in montane to alpine conditions on ultramafic ground.

Propagation difficult

This is one of New Zealand's most endangered plants and is confined to a 6 ha area of limestone debris at Castle Hill in Canterbury. Propagation is difficult and seedlings are rarely seen; seed set is poor and germination is difficult (McCaskill 1982). Key threats are browsing animals and weeds (NZPCN n.d. [c]). It thrives best when the crown is buried under the soil debris (McCaskill 1982).

Nigella no friends — one female only

Endemic to Three Kings islands, only one tree (female) was ever found in the wild. However, some fruit were produced in 1989 and these were harvested. If there are no male trees, how is pollination achieved? Is pollen arriving from the mainland or is the plant capable of self-pollination? In cultivation, fruits form but all the seeds are hybrids with the common species *P. corymbosa*. While *P. baylisiana* can be grown from stem cuttings or from basal suckers, these can take 10 months to strike (grow roots) and are liable to complete collapse (Oates n.d.).

Clianthus maximus, kākā beak.

Zealand species. *Pennantia baylisiana* (see the box on page 186) is alive today because it has been propagated vegetatively in cultivation. There is still a problem, though. Anything produced by vegetative propagation is a clone of the original — so while the species is technically alive, it is also functionally extinct because there is only one genetic individual, a female plant.

A similar example is the kākā beak (*Clianthus maximus*), which is hanging on (literally — it is often found on cliff faces) in the wild, but is quite common in garden cultivation. Could these garden plants be used to re-establish populations in the wild? A study of genetic diversity found that the garden cultivars were not representative of the diversity of the wild plants (Song et al. 2008);

research such as this helps us to avoid the potential error of putting the 'wrong' plant material back into the wild.

Another form of ex situ conservation of plants is to collect seeds and store them in a seed bank. The collected seeds are cleaned and dried, then stored at −20°C and 15 per cent relative humidity: under these conditions many seeds will survive for many years. Two of the most significant seed banks globally are the Millennium Seed Bank at Kew Gardens in Britain, which contains 2.3 billion seeds of 40,000 species (RBG Kew n.d. [a]), and the vault at Svalbard in Norway which contains 968,000 seed samples (Crop Trust n.d.). New Zealand has one seed bank, the Margot Forde Forage Germplasm Centre at

Winika cunninghamii, one of the New Zealand orchids with seeds that can't presently be stored in seed banks.

AgResearch in Palmerston North, which holds 65,000 samples of crop and crop-related species (grass, clover and other crop plants) that are the backbone of New Zealand agriculture. Since 2013 the Margot Forde Forage Germplasm Centre has also been the facility for the New Zealand Indigenous Flora Seed Bank (NZIFSB) (McGill et al. 2018).

Underpinning the NZIFSB is the question of which species should take priority for collection; at first glance it would seem that the At Risk or Threatened species would be the obvious focus.

An algorithm used in 2017 to analyse a range of factors determined that the Castle Hill buttercup is New Zealand's most threatened plant (Mitchell 2018). A small quantity of seed has already been collected and banked, but the species is not yet secure as it hardly ever produces any seed. Why is that? One reason is that mice eat the seeds and hares eat the plants (NZPCN n.d. [c]; Radio New Zealand 2017), but currently we do not know whether other factors are also involved. This highlights the importance of getting the species into cultivation so that we can study its behaviour and figure out how to grow and propagate it.

However, seed banking should not be limited to rare species; keystone species are also important. For example, pōhutukawa (*Metrosideros excelsa*) is not rare, but it is a key ecological and structural component of some forest types and the recent arrival in New Zealand of the disease myrtle rust, which has the potential to destroy pōhutukawa forests, means that pōhutukawa is now severely threatened (de Lange et al. 2018). Such keystone species are a high priority for seed banking, and since the arrival of the disease 500 collections (14.5 million seeds) of pōhutukawa and its relatives in the myrtle family have been lodged in the NZIFSB (McGill et al. 2018).

As long as it is possible to collect the seed in the first place, seed banking is an excellent form of ex situ conservation; however, it doesn't work for many New Zealand species because their seeds do not survive the drying process. Maire tawake (*Syzygium maire*), a large tree from the myrtle family, has a fleshy seed which dies when dried for seed-bank storage (van de Walt 2018), yet this species is of high priority for conservation because it is in the myrtle family and will be affected by myrtle rust. Scientists at Massey

University are experimenting with cryo-storage, whereby the embryo is extracted from the seed, encapsulated in a protective coating and stored in liquid nitrogen (van de Walt 2018). While the basic process has been around for several years, we do not yet know how to apply it to different species that have different seed characteristics.

New Zealand orchids (177 species of which 31 are At Risk and 11 are Threatened; NZPCN n.d. [b]) present another problem — their tiny seeds have little flesh to protect the embryo or support the seed in storage, and for many species traditional seed banking fails. This problem is also found globally, where many orchid species are threatened yet cannot be seed-banked at present (Lee & Yeung 2018). In a New Zealand project, scientists are investigating orchid seeds to try to understand their morphology and physiology and figure out a way to store them successfully in a seed bank.

In an interesting twist, another form of ex situ conservation in New Zealand involves the conservation of exotic species. Alongside our own approximately 3000 native plant species there are some 40,000 exotic cultivated plant species in New Zealand (Dawson 2010). (Yes, more than 10 times the number of native plant species.) This huge range of plants has been brought to New Zealand ever since people began coming to these shores. Māori were the first to do so when they brought taewa (Māori potato) to New Zealand. This group of plants is now considered a taonga and is the subject of a conservation programme to collect and retain the different forms of taewa in cultivation (Roskruge 2014).

European settlers also brought exotic plants to New Zealand, but in greater volume, bringing plants for pasture, forestry, food and decoration. Today we have strict biosecurity regulations in New Zealand, and plants can only be imported under certain conditions; however, in the early days regulation was less strict and thousands of species were imported. Among the plant species brought to New Zealand, are hundreds that are now rare or threatened in their home habitat, making cultivated collections the only remaining way to ensure the survival of some of these species. Horticultural skills are critical for some of these rare species as they are hard to propagate and to grow, which is partly why they are rare in the first place.

Figuring out how to propagate and grow rare plants in ex situ collections is only one of the many horticultural challenges associated with conservation, whether it is for New Zealand native flora or for exotic species. Propagation, growing media, flowering, seed production, pests and diseases — these are all part of the horticulture of conservation.

Urban horticulture

In production horticulture we deliberately choose the best sites on which to grow a particular crop, then apply the best combination of water and nutrients to achieve the best possible growth of the crop. In urban horticulture, and particularly for street trees, we seldom have this luxury. Many street-tree planting sites are characterised by poor or no soil, low fertility or even contamination from construction of roads or nearby buildings, severe compaction such that roots struggle to grow, lack of water because the surface has been paved over, small planting spaces, and unhelpful microclimates like all-day shade and wind tunnels created by large buildings. Yet we expect trees to grow in these places! Problems such as these suggest that it is too difficult to try to grow trees in these circumstances; however, urban vegetation is vital for both human and environmental health.

Evidence now shows that interacting with, or being exposed to, nature is beneficial to health (Dover 2015; Kaplan S. 1992, 1995; Thompson 2011; Ulrich 1999). In addition, 'nature' can be as broad as a vast wilderness or as simple as a row of pot plants on a windowsill (Kaplan R. 1992). The research also shows that when given a choice, most people prefer situations that contain plants and this preference operates at an instinctive level (Kaplan S. 1995; Timm, Dearborn, and Pomeroy 2018; Ulrich 1999). For

most people, and particularly urban people, the opportunity to visit parks and gardens, walk among trees or live in a neighbourhood with street trees has positive health outcomes (Wood et al. 2018). It works indoors, too: offices and homes in high-rise buildings that contain living plants have better atmospheric conditions and are better working environments (Dover 2015).

When it comes to the outdoor environment, cities are hot and dry 'heat islands' (because of the amount of hard-surface they contain), often with poor soil, frequent contamination and pollution, and modified water processes. For example, much rainwater does not percolate into the soil because most of the soil is paved over, so stormwater run-off accumulates and causes flooding. Urban vege-

Urban plantings in Singapore.

tation, particularly large trees, can mitigate some of those problems. Through normal biological processes, urban vegetation can reduce pollution, reduce temperature, increase humidity, and reduce water run-off (Bliss, Newfold, and Rias 2009; Chen and Jim 2008; McBride 2017; Rogers et al. 2017). While knowledge of these functions is not new (e.g. Robinette 1972), it is only in recent times that their importance in improving the urban environment has been widely appreciated. Indeed, cities and urban areas are now recognised as hotspots for biodiversity (Lawton 2019; Shaffer 2018), with that biodiversity being essential to the sustainability of cities and the quality of life they provide (de Lurdes Chorro Barrico et al. 2017; Wood et al. 2018).

The urban horticulturalist has to figure out how to grow plants in often hostile circumstances where the normal rules of cultivation may not apply. The first key lesson is that because of the harsh conditions, urban trees are shorter-lived than the same species planted in a field or a forest (Close, Nguyen, and Kielbaso 1996; Quigley 2004). A second aspect is that some species, such as the London plane tree (*Platanus × acerifolia*), which can withstand street conditions, are too large for the constrained space available in most urban areas. Plant selection is therefore a key issue. Several researchers have tested different species and cultivars to find out how they perform and survive in street plantings (Arnold et al. 2012; Koeser et al. 2013; Lawrence et al. 2012; Percival, Keary, and Al-Habsi 2006; Rhoads, Meyer, and Sanfelippo 1982; Sjöman, Hirons, and Bassuk 2015), with the usual objective of selecting superior forms.

You might think that the solution for street planting is to grow local native plants; however, this approach frequently fails. While local species are well adapted to local climate conditions,

street plantings often have to endure severe substrate, water and microclimate conditions that differ considerably from the native habitat of the species and even local plants cannot survive those extremes. Another approach is to use environmental analogues; for example, if the street situation is high altitude and cold, choose plants that come from similar conditions. Various studies have been done on this kind of matching (Widrlechner 1994; 2004). While this approach can be useful, it does not take sufficient account of the factors of adaptability and obligation. An example of obligation is the above-mentioned herb *Celmisia spedenii* that is probably an obligate ultramafic species — that is, it must have ultramafic substrate to grow successfully. In contrast, other species may grow perfectly well in a range of different circumstances; that is, they have the capacity to adapt.

The ability to adapt, or not, cannot be determined from physical appearance — we have to grow the plant and find out what it will do, and sometimes the results can be interesting. Two trees on Massey University's Palmerston North campus illustrate the point. *Magnolia campbellii* is from the cool temperate zone in the Himalayas, and is supposed to live about 100 years and grow to 30 m high in its native habitat. *Populus deltoides* is from swampy ground in the southern states of the United States, and is also supposed to live about 100 years and grow to about 30 m tall in its native habitat. The plants on campus are both about 60 years old. Field observation shows that the *Magnolia* is only 10 m tall and has stopped growing; it will never reach its expected 30 m. The *Populus*, nearby, is 29 m tall and is still growing, so potentially will get larger still. Why do these differences occur? Finding out is one of the interesting topics of urban horticulture.

Parks and gardens

Parks and gardens are more horticulturally friendly than street-tree situations. In parks and gardens the soil fertility and water factors are usually much better than for street plantings, and horticulturalists have better control over growing conditions. Parks and gardens come in all shapes and sizes — indoors or outdoors, urban or rural, small or large, public or private sites — and sometimes have horticultural themes like rose gardens or native plants. A typical public park in New Zealand will have elements such as a rose garden, a display glasshouse, a native plant area and various types of flower display gardens. A key aspect of parks and gardens is the visual and artistic qualities of the plants, which can involve anything from the artistic composition of a group of plants to the design of whole gardens. As well as the health and environmental benefits, parks and gardens have the added human dimensions of aesthetic, cultural and historical values.

An excellent example of these values is the Hampton Court Palace gardens in London. Originally created as a pleasure garden in the 1500s and still a royal park today, the grounds have vast significance for the events that took place there but are also an example of certain design styles. The Pond Gardens were originally a fish farm in medieval times, but were converted by William and Mary (circa 1690–1700) to house her collection of plants from around the world. It was restored in the 1990s and, as the image shows, requires precise horticulture to get the lawns, topiary, and flower displays just right (Hampton

The Pond Gardens at Hampton Court Palace, London.

The Indian Char Bargh Garden at Hamilton Gardens.

Court's 60 acres of gardens are managed by 38 staff and the on-site nursery grows 140,000 plants each year [Hampton Court Palace Garden and Estate n.d.]).

In New Zealand we do not have decorative gardens dating from the fifteenth century, although we do have examples of early food-production sites; however, we have public parks that illustrate the artistic aspects of horticulture and the use of plants for visual and design displays. An excellent example is Hamilton Gardens. A series of more than 20 themed gardens illustrate design themes, different eras or cultural styles such as Italian, American, Japanese and Chinese cultures (Hamilton Gardens n.d.). These gardens are now a major tourist attraction for Hamilton and attract around one million visitors per year.

A key horticultural challenge for parks and gardens is to grow a range of plants and get them all to grow and flower in the right way at the right time. While production horticulturalists have to know absolutely everything about the crops they grow, say apples and pears, the number of crops is relatively limited; whereas the horticulturalist who manages something like Hamilton Gardens has to know the requirements of many species.

Another example of this kind of horticultural challenge is the growing of tropical plants for the conservatory display houses found in most town or city parks. Some of these display houses contain a cornucopia of plant life forms from many different origins. Not only does the horticulturalist need to know the requirements of the many species, but they also have to figure out how to use one greenhouse, with one set of climatic conditions, to grow all those species successfully.

One of the great fascinations of horticulture is the huge diversity of form and growth of plants. Some great examples can be seen in the display glasshouse at Victoria Esplanade in Palmerston North.

Aristolochia gigantea, commonly known as the Dutchman's pipe because the bud is shaped

Aristolochia gigantea, Dutchman's pipe.

Tacca chantrieri, cat's whiskers or black bat flower.

like a pipe, is a climber from Panama and Brazil. It has huge flowers, the size of a tea towel, yet those flowers arise from thin, delicate-looking stems. Even though the vine looks delicate, if it is not severely pruned each year it would take over the whole glasshouse.

The cat's whiskers (*Tacca chantrieri*), also known as the black bat flower, is a species that grows on the forest floor in Asia. Some horticulturalists thought that it had potential as a commercial flowering pot plant (because of its unique appearance and colour), but found that it was hard to propagate because most of the seeds had no embryo (Krisantini, Wiendi and Palupi 2017). In addition, its value as a flowering pot plant depends on the presence of the flower, so research was needed to figure out how to get it to flower out of season (for year-round supply).

As well as unusual flowers, many species have interesting growth habits. Wendland Cape primrose (*Streptocarpus wendlandii*) is a ground-dwelling plant from South Africa that just grows one big leaf — and in fact this is not even a true leaf (it is termed a phyllomorph). *Streptocarpus* is a dicot (a plant that has two cotyledon leaves), but something unusual happens in the meristem (the growing-point tissue where cell division takes place). Instead of growing a crown of many leaves like most plants do, one cotyledon overwhelms the other and the dominant one grows and grows until it is 40–50 cm long and looks like a leaf (Nishii et al. 2015). A flower stalk arises from the base of the leaf. Having just one leaf might seem to be a risky strategy, but this plant seeds prolifically and the horticulturalist has to remove seedlings to stop the species spreading too far.

Air plants (*Tillandsia* spp.) also rely on their leaves, because they don't have any roots. The functions of the root system are to hold the plant in place and to provide it with water and nutrients — *Tillandsia* achieves these functions with specialised scales (trichomes) on its stems and leaves that absorb water and nutrients (Kew Science n.d.). It does not need to be held up because its stems drape over branches and hang down — successful cultivation requires a branch for it to grow on and good humidity. Pots and potting mix are not required.

New Zealand species are not generally found in greenhouse displays because our flora is not tropical; however, native species are used in many ways in the outdoor spaces of parks and gardens, such as New Zealand ferns, orchids and foliage plants being used in a coolhouse display.

Streptocarpus wendlandii, Wendland Cape primrose.

Another dimension of parks and gardens is the private garden, where people cultivate plants for home use. In New Zealand, cultivation for domestic consumption first focused on fruits and vegetables. In the past a vast range of such plants was offered for sale; a search of old nursery catalogues reveals an abundance of types. Most of these old fruits and vegetables have disappeared from cultivation today; many were not suited to commercial production and some simply went out of fashion. Today, some people are committed to preserving these old types; the taewa project referred to earlier is one example, while heritage fruits and vegetables is another.

Home gardeners and plant enthusiasts also grow an enormous range of plants for aesthetic and botanical interest. This includes trees, flowering shrubs, climbing plants, flowering perennials and annuals, bulbs, hothouse plants, orchids, alpine plants, rhododendrons and roses, to name a few. Recall the approximately 40,000 exotic plant species that are present in New Zealand — some of these are the food and

Air plant (species of *Tillandsia*).

Hippeastrum aulicum, a frost-tender bulb from Brazil and Paraguay, is one of many thousands of cultivated ornamental horticultural plants present in New Zealand.

(the website is listed at the end of the chapter). New Zealand does not yet have a comprehensive documentation of cultivated plants; the Royal New Zealand Institute of Horticulture records the existence of collections but not the detail of species in those collections. Sometimes interesting discoveries are made, such as a gladiolus species that was thought to be extinct in Britain but which was found in Canterbury (Jamieson 2015), and two historic daffodils (bred in 1629 and 1768) that were also found in gardens in Canterbury (Rooney 2014).

For the plant enthusiast there are many interesting societies and interest groups both nationally and internationally. Trees are the focus of the International Dendrology Society, the Global Trees Campaign and the International Society of Arboriculture. There are specialist groups for rhododendrons (New Zealand Rhododendron Association), orchids (Orchid Council of New Zealand), alpine plants (New Zealand Alpine Garden Society), and many more. Plants and gardens in general are covered by the Royal Horticultural Society in Britain, Plant Heritage (mentioned previously), the New Zealand Gardens Trust and many local horticultural societies in New Zealand. As well as the botanic gardens, some key New Zealand sites are Eastwoodhill Arboretum near Gisborne and Pukeiti in Taranaki.

forage species that enable agriculture and horticulture, and some are weeds (about 300 species are invasive weeds), but the vast majority of the 40,000 are cultivated ornamental plants grown for decorative purposes.

Although most of these decorative plants are not wild species (they are often selections or hybrids that were developed for their decorative features), many have conservation value because, as with heritage fruits and vegetables, they represent horticultural history and encompass useful genetic variation.

One interesting society is Plant Heritage, which encompasses sites in England, Ireland and the Channel Islands, and aspires to document and preserve cultivated plants in Britain

Cut-flower production

Another example of non-food horticulture is the growing of flowers and foliage as cut products for indoor or outdoor display. New Zealand exported $169 million worth of plants, flowers and seeds in

2018 (Plant and Food Research 2018), a relatively small quantity compared with the $3086 million worth of fresh fruit and vegetables exported that year. Exports of flowers, plants and seeds included $20.2 million worth of cut flowers, primarily sent to Asia, and $42.4 million worth of bulbs, corms and tubers, primarily lilies and tulips (Plant and Food Research 2018). The main flower exports in 2018 were orchids ($11.5 million), paeonies ($3.3 million) and hydrangeas ($1.6 million); of several other species exported, none has an export value of more than $1 million (Plant and Food Research 2018). Production for export is dominated by eight main growers who produce 95 per cent of the flowers exported from New Zealand (New Zealand Flower Exporters Association 2018). Another aspect of the flower industry is the growth and sale of the bulbs, corms or tubers from which the flowers are grown. In 2018 New Zealand exported $26 million worth of lily bulbs and $15.8 million worth of tulip bulbs (Plant and Food Research 2018).

In 2017 there were 330 floriculture businesses in New Zealand, with Nelson, Auckland and Waikato the key regions (Statistics New Zealand 2017). Undercover cultivation of 105 ha and outdoor cultivation of 1762 ha represented an investment of $435 million (Plant and Food Research 2018). Although most New Zealand businesses are small and New Zealand is a small market, internationally the growing and sale of plants and flowers can generate considerable returns.

In this form of horticulture, flowers and foliage are grown, harvested and sold for use in floral bouquets and decorations. Florist shops in New Zealand carry a wide range of flowers, including roses, carnations, chrysanthemums, lilies, paeonies, calla lilies, daffodils, *Gypsophila*,

The first New Zealand flower growers

The earliest growers of flowers for decorative purposes in New Zealand were Māori who cultivated kōwhai ngutukākā (kākā beak, *Clianthus puniceus* and *Clianthus maximus*) at kāinga and pā sites around the Kaipara Harbour and East Cape.

Alstroemeria, tulips and foliage such as fern, eucalyptus and palm leaves. Some of these products are imported and some are obtained from local growers who focus on the domestic market.

While some flower production takes place outdoors, most flowers are grown under cover in some form of greenhouse or protected growing environment. Greenhouses protect the develop-

Freesias in a greenhouse at Wilflora in the Manawatū.

ing crop from wind and rain damage, and also allow growers to manipulate soil, temperature, fertiliser, water, and light and day-length as they manage the production process.

For example, at Wilflora in the Manawatū, freesias are grown under cover and the grower is able to manipulate several factors:

- Temperature is controlled by a combination of fans plus lifting or lowering the ends of the greenhouse.
- Water is provided through drip irrigators.
- Fertiliser is applied to the soil before the crop

is planted and also as the crop grows.

The wire structures over each row support the crop and hold the flowers upright as they form.

How does the grower know what kind of growing environment to provide? Many flower species, such as freesias, require a period of chilling to trigger flower formation, while for others (such as chrysanthemums) the stimulus is changing day-length. Growers manipulate these factors to lengthen the production period and increase production of the particular flower.

Winter in summer — no problem!

The freesia is a South African flower where the bulb needs to receive a certain amount of chilling to trigger flower formation. Normally the bulb would receive this chilling over the winter with the flower being produced in spring. But this short flowering season is not enough to support a viable business, and out-of-season flowers command high prices in the market. Clever growers have figured out how to fool the plant into producing flowers at other times of the year. At Wilflora they use a network of buried pipes carrying chilled water to chill the soil and bulb and provide the necessary cold, thereby allowing the grower to produce flowers for most of the year.

Rhododendron Highland White Jade.

Successful production involves manipulating the environmental conditions to provide the necessary triggers for growth and flowering.

Plant production and the nursery industry

For all forms of horticulture, plant selection and breeding, propagation and plant production are key elements.

Plant selection has two broad approaches depending on the end application. For conservation purposes two key principles are important. First, any plants grown in an ex situ living collection (such as a botanic garden) should be wild-sourced — that is, the original seeds or cutting material should have been obtained from a wild population in its natural habitat — as the purpose of ex situ conservation is to conserve the natural genetic diversity of the species. Second, plants should be eco-sourced — if, say, the end use is a revegetation project, the plants used should ideally be sourced from the local area, as plants sourced from other regions may not represent the local type.

A different approach to plant selection is used for cultivated horticultural plants, where superior forms are sought and propagated for a particular use. Some forms arise as a chance mutation on an existing plant (some apple cultivars arose in this fashion), while in other cases superior seedlings arise randomly in a batch of seedlings. Systematic breeding is also used. Appropriate parent plants are selected, a hybrid cross is made and the best seedlings examined for their horticultural qualities. In this manner horticulturalists breed lines of plants that are

Plant diversity

Plant diversity is the resource that horticulturalists use to identify new possibilities and then breed and select new lines of plants. Production horticulturalists emphasise factors like fruit size and colour, yield, keeping quality, and ability to withstand mechanical harvesting. Parks and gardens horticulturalists look for good growth characteristics (e.g. pest and disease resistance, frost tolerance) and good aesthetic features (e.g. long period of flower display, double flowers, weather-hardy flowers, on-trend colours). Often breeders look for unusual features such as black flowers in tulips and roses, while orange flowers in *Cymbidium* orchids is the aim of one New Zealand breeder.

disease-resistant, cold-tolerant, longer-flowering, or whatever quality is desirable at the time. Hybrid crosses were made, for example, to combine the good visual quality of one parent with the disease resistance of the other parent.

In some cases the crosses are quite complex. To achieve *Rhododendron* Highland White Jade (shown on page 199), four parents were involved; its lineage is (Dr Herman Sleumer × *R. herzogii*) × (*R. laetum* × *R. aurigeranum*).

While traditional breeding relied on physical observation of morphology and growth characteristics, today breeders use molecular screening of potential parents or lines of seedlings. This allows them to quickly identify desirable characteristics and eliminates the lengthy wait to find out how a line will behave in the field.

Once desirable plants are obtained, propagation and production are required. Plant propagation is done in two main ways: sexual propagation (by seed), and asexual propagation where vegetative material (usually stems and leaves) is used to generate a new plant.

Seed propagation is useful where large quantities of plants are needed and where variation (inherent in any population of seedlings) is acceptable, or in fact desirable in the case of native plants for conservation projects.

Vegetative propagation is used where variation is not wanted, and is essential where a cultivar is involved (e.g. many rhododendrons and roses) as vegetative propagation must be used to perpetuate the features of the cultivar (cultivars do not 'come true' from seed). Similarly, in production horticulture all fruit cultivars are propagated vegetatively, usually by budding or grafting.

The basic theory of propagation is well known; however, the application is often less easy and there are still many plants that are difficult to propagate. The usual methods fail and we do not always know why — more research is needed on many aspects of propagation.

Plant propagation is the role of the nursery industry. In 2017 there were 159 undercover and 369 outdoor nursery production enterprises in New Zealand, with Auckland, Nelson, Waikato and Bay of Plenty being key regions (Statistics New Zealand 2017). Most enterprises are less than 20 ha in size (Statistics New Zealand 2017) and have an average of 3.5 full-time staff (Westpac Institutional Bank 2016). Some nurseries specialise in particular lines, such as roses, rhododendrons, native plants or forestry species. The nursery industry is represented by New Zealand Plant Producers Incorporated (NZPPI),

Forms of plant propagation. From left pictures a–c show grafting where the desirable scion (top portion) is fused onto a rootstock (the lower portion). This method is commonly used to confer disease resistance and to manipulate the size of the plant, as the characteristics of the rootstock are conveyed into the scion. Pictures d and e show vegetative propagation by stem cuttings and leaf cuttings, respectively. Picture f shows an orchid that develops plantlets (with roots) among the foliage, which can be used to grow new plants.

which furthers the interests of this industry and promotes those aspects of horticulture that are associated with plant propagation.

The future

If not food then what? The food production aspects of horticulture are all about direct values — the generation of food, and of shelter (in the case of forestry). However, equally important are the indirect values, which include the life-supporting functions of the living environment and the human values like history, culture, society and beauty. We do not always exchange money for, say, the historical value of an amazing garden or the health value of a city park, but nevertheless these functions are hugely valuable and this value is becoming much better understood.

In future two components are likely to come to the fore. First will be urban horticulture

A roof garden at Singapore Botanic Garden.

— street trees, green mosaics and linkages, urban parks and green spaces, town plantings — a green infrastructure that utilises plants to generate ecosystem service functions and support human health. In addition to the traditional street and park green spaces, applications such as green roofs and walls will become much more common, although we still have to figure out some of the engineering and horticultural issues before they can become standard practice.

The second component should be a more comprehensive recognition of the importance of biodiversity as the resource that underpins all forms of horticulture (and indeed most aspects of life [da Cunha, Mace, and Mooney 2019]), and a greater understanding of that diversity and the work required for its conservation. 'Knowing your plants' is a fundamental part of horticul-

ture and with some 400,000 species of plants worldwide there are endless opportunities and much for the horticulturalist to do and learn!

Further information

Professional and industry organisations for horticulturalists:

- Botanic Gardens Australia and New Zealand Incorporated (www.bganz.org.au/)
- Botanic Gardens Conservation International (UK) (www.bgci.org/)
- New Zealand Plant Conservation Network Rōpū hononga Koiora Taiao ki Aotearoa (www.nzpcn.org.nz/)
- New Zealand Plant Producers Incorporated (https://nzppi.co.nz/)

- International Plant Propagators Society (http://nz.ipps.org/)
- Royal Horticultural Society (UK) (www.rhs.org.uk/)
- Royal New Zealand Institute of Horticulture (www.rnzih.org.nz/)

New Zealand horticultural journals:
- *Commercial Horticulture Journal* (www.nursery.net.nz/shop/category.aspx?catid=1)
- *New Zealand Garden Journal* (www.rnzih.org.nz/pages/NZ_Garden_Journal_index.htm)
- *New Zealand Rhododendron Journal*

Horticultural societies and public gardens:
- Eastwoodhill, National Arboretum of New Zealand (www.eastwoodhill.org.nz/)
- Global Trees Campaign (https://globaltrees.org/)
- International Dendrology Society (http://dendrology.org/)
- International Society of Arboriculture (www.isa-arbor.com/)
- New Zealand Alpine Garden Society (www.nzags.com/)
- New Zealand Gardens Trust (www.gardens.org.nz/)
- New Zealand Rhododendron Association (www.rhododendron.org.nz/)
- Orchid Council of New Zealand (www.orchidcouncil.org.nz/)
- Plant Heritage, UK (www.plantheritage.org.uk/)
- Pukeiti (www.trc.govt.nz/gardens/pukeiti/)

References

Arnold, M.A., Bryan, D.L., Cabrera, R.I., Denny, G.C., Griffin, J.J., Iles, J.K., King, A.R., et al. 2012. 'Provenance experiments with baldcypress, live oak, and sycamore illustrate the potential for selecting more sustainable urban trees.' *Arboriculture & Urban Forestry* 38 (5): 205–13.

Bliss, D.J., Newfeld, R.D., and Ries, R.J. 2009. Storm water runoff mitigation using a green roof. *Environmental Engineering Science* 26 (2): 407–17.

Botanic Gardens Conservation International. n.d. [a]. 'About plant conservation.' Accessed 1 August 2019. www.bgci.org/about/about-plant-conservation.

Botanic Gardens Conservation International. n.d. [b]. 'Convention on Biodiversity.' Accessed 1 August 2019. www.bgci.org/our-work/policy-and-advocacy/convention-on-biological-diversity.

Botanic Gardens Conservation International. n.d. [c]. 'Global Strategy for Plant Conservation.' Accessed 1 August 2019. www.bgci.org/our-work/policy-and-advocacy/the-global-strategy-for-plant-conservation.

Chen, W.Y., and Jim C.Y. 2008. 'Assessment and valuation of the ecosystem services provided by urban forests.' In *Ecology, planning and management of urban forests. International perspectives,* edited by Margaret M. Carreiro, Yong-Chang Song, and Jiangguo Wu, 53–83. New York: Springer.

Close, R.E., Nguyen, P.V., and Kielbaso, J.J. 1996. 'Urban versus natural sugar maple growth: I. Stress symptoms and phenology in relation to site characteristics.' *Journal of Arboriculture* 22 (3): 144–50.

Crop Trust. n.d. 'Svalbard Global Seed Vault.' Accessed 1 August 2019. www.croptrust.org/our-work/svalbard-global-seed-vault.

da Cunha, M.C., Mace, G.M., Mooney, H. (eds.). 2019. 'Report on the Plenary of the Intergovernmental Science-Policy Platform on Biodiversity and Ecosystem Services on the work of its seventh session.' Annex: Summary for policymakers of the global assessment report on biodiversity and ecosystem services of the Intergovernmental Science-Policy Platform on Biodiversity and Ecosystem Services. https://ipbes.net/global-assessment.

Dawson, M.I. (ed.) 2010. 'Documenting New Zealand's cultivated flora: "A supermarket with no stock inventory".' Report from a TFBIS-funded workshop held in Wellington, New Zealand, 9th September 2009. Version 2. Landcare Research. Accessed 2 December 2019. www.landcareresearch.co.nz/publications/researchpubs/Report-documenting_New_Zealands_cultivated_flora.pdf.

de Lange, P.J., Rolfe, J.R., Barkla, J.W., Courtney, S.P., Champion, P.D., Perrie, L.R., Beadel, S.M., et al. 2018. Conservation status of New Zealand indigenous vascular plants, 2017. Wellington: Department of Conservation. Accessed 16 November 2018. www.doc.govt.nz/Documents/science-and-technical/nztcs22entire.pdf.

de Lurdes Chorro Barrico, M., Castro, H., Coutinho, A., Silva Gonçalves, M.T., Freitas, H., and de Oliveira Castro, P.C. 2017. 'Plant and microbial biodiversity in urban forests and public gardens: Insights for cities' sustainable development.' *Urban Forestry and Urban Greening* 29: 19–27.

Dover, J.W. 2015. *Green infrastructure: Incorporating plants and enhancing biodiversity in buildings and urban environments.* Oxford: Routledge.

Hamilton Gardens. n.d. Accessed 20 February 2020. https://hamiltongardens.co.nz/collections.

Hampton Court Palace Garden and Estate. n.d. Accessed 27 February 2020. www.destinations-uk.com/articles.php?country=england&id=69&articletitle=Hampton%20Court%20Palace%20Gardens%20and%20Estate.

Jamieson, C. 2015. '*Gladiolus* × *brenchleyensis*: A near-extinct relict of Great Britain is rediscovered in New Zealand.' *New Zealand Garden Journal* 18 (1): 2–8.

Kaplan, R. 1992. 'The restorative benefits of nearby nature.' In *The role of horticulture in human well-being and social development*, edited by Diane Relf, 123–33. Portland, OR: Timber Press.

Kaplan, S. 1992. 'The restorative environment: nature and human experience.' In *The role of horticulture in human well-being and social development,* edited by Diane Relf, 134–45. Portland, OR: Timber Press.

Kaplan, S. 1995. 'The restorative benefits of nature: Toward an integrative framework.' *Journal of Environmental Psychology* 15 (3): 169–82.

Kew Science. n.d. 'Plants of the World *online*.' Online at Accessed 27 February 2020. http://powo.science.kew.org.

Koeser, A., Hauer, R., Norris, K., and Krouse, R. 2013. 'Factors influencing long-term street tree survival in Milwaukee, WI, USA.' *Urban Forestry & Urban Greening* 12 (4): 562–8.

Krisantini, W., Ni Made Arment, P., Endah Retno, P.. 2017. 'Evaluation of horticultural traits and seed germination of *Tacca chantrieri* 'André'.' *Agriculture and Natural Resources* 51 (3): 169–72.

Lawrence, A.B., Escobedo, F.J., Staudhammer, C.L., and Zipperer, W. 2012. 'Analyzing growth and mortality in a subtropical urban forest ecosystem.' *Landscape and Urban Planning* 104 (1): 85–94.

Lawton, G. 2019. 'All hail the urban jungle.' *New Scientist* 243 (3239): 24.

Lee, W.G., and Reeves, R.D. 1989. 'Growth and chemical composition of *Celmisia spedenii,* an ultramafic endemic, and *Celmisia markii* on ultramafic soil and garden loam.' *New Zealand Journal of Botany* 27 (4): 595–8. https://doi.org/10.1080/0028825X.1989.10414144.

Lee, Y.I., and Yeung, E.C. (eds.). 2018. *Orchid propagation: from laboratories to greenhouses – methods and protocols.* New York: Humana Press, Springer Nature.

McBride, J. 2017. *The world's urban forests: History, composition, design, function and management.* Future City series No. 8. Switzerland: Springer International.

McCaskill, L.W. 1982. 'The Castle Hill buttercup (*Ranunculus paucifolius*): A story of preservation.' Canterbury: Lincoln College, Tussock Grasslands & Mountain Lands Institute.

McGill, C., Nolan, K., Schnell, J., MacKay, M., Winkworth, C., Aubia, G., and McGlynn, V. 2018. 'Seed banking (conservation) in New Zealand: supporting *in situ* conservation.' *IPPS Proceedings* 68: (in press).

Mitchell, C. 2018. 'The ark, the algorithm and our conservation conundrum.' *Stuff,* 12 July 2018. Accessed 16 November 2018. www.stuff.co.nz/environment/103568532/the-ark-the-algorithm-and-our-conservation-conundrum.

New Zealand Flower Exporters Association. Accessed 20 January 2019. www.nzflowers.com/home.asp.

New Zealand Plant Conservation Network. n.d. [a]. '*Celmisia spedenii.*' Accessed 1 August 2019. www.nzpcn.org.nz/flora_details.aspx?ID=432.

New Zealand Plant Conservation Network. n.d. [c]. '*Ranunculus pauciflorus.*' Accessed 1 August 2019. www.nzpcn.org.nz/flora_details.aspx?ID=2372.

New Zealand Plant Conservation Network. n.d. [b]. New Zealand Orchidaceae (search results). Accessed 1 August 2019. www.nzpcn.org.nz/flora_search.aspx?scfSubmit=1&scfStart_Results=0&scfFlora_Category=&scfLatin_

Name=&scfCommon_Name=&scfFamily_
Name=orchidaceae&scfDistrict=&scfAncestry=
&scfFlora_Structural_Class=&scfFlora_
Status=&scfNative_Or_Exotic=1.

Nishii, K., Hughes, M., Briggs, M., Haston, E., Christie, F., DeVilliers, M.J., Hanekom, T., Roos, W.G., Bellstedt, D.U., and Möller, M. 2015. '*Streptocarpus* redefined to include all Afro-Malagasy Gesneriaceae: Molecular phylogenies prove congruent with geographical distribution and basic chromosome numbers and uncover remarkable morphological homoplasies.' *Taxon* 64 (6): 1243–74. http://dx.doi.org/10.12705/646.8.

Oates, M. n.d. Cultivation of *Pennantia baylisiana*. Accessed 27 February 2020. www.rnzih.org.nz/pages/pennantia.htm.

Percival, G.C., Keary, I.P., and Al-Habsi, S. 2006. 'An assessment of the drought tolerance of *Fraxinus* genotypes for urban landscape plantings.' *Urban Forestry & Urban Greening* 5 (1): 17–27.

Plant and Food Research. 2018. 'Fresh facts: New Zealand horticulture.' Accessed 27 February 2020. www. freshfacts.co.nz/files/freshfacts-2018.pdf.

Quigley, M.F. 2004. 'Street trees and rural conspecifics: Will long-lived trees reach full size in urban conditions?' *Urban Ecosystems* 7: 29–39.

Radio New Zealand. 2017. 'Critter of the week: The Castle Hill buttercup.' On *Afternoons with Jessie Mulligan*. Accessed August 1, 2019. https://www.rnz.co.nz/national/programmes/afternoons/audio/201842819/critter-of-the-week-the-castle-hill-buttercup.

Rhoads, A.F., Meyer, P.W., and Sanfelippo, R. 1981. 'Performance of urban street trees evaluated.' *Journal of Arboriculture* 7 (5):127–32.

Robinette, G.O. 1972. *Plants, people and environmental quality: A study of plants and their environmental functions*. US National Parks Service.

Rogers, K., Andreucci, M-B., Jones, N., Japelj, A., and Vranic, P. 2017. 'The value of valuing: recognising the benefits of the urban forest.' In *The urban forest: Cultivating green infrastructure for people and the environment* (Future City series No. 7), edited by David Pearlmutter, Carlo Calfapietra, Roeland Samson, Liz O'Brien, Silvija Krajter Ostoić, Giovanni Sanesi, and Rocio Alonso del Amo, 283–99. New York: Springer.

Rooney, D. 2014. 'Daffodils from the past.' *New Zealand Garden Journal* 17 (2): 11–13.

Roskruge, N. 2014. *Rauwaru, the proverbial garden: Ngā-Weri, Māori root vegetables, their history and tips on their use*. Institute of Agriculture and Environment, Massey University.

Royal Botanic Garden Kew. n.d. [b]. 'Millennium Seed Bank.' Accessed 1 August 2019. www.kew.org/wakehurst/whats-at-wakehurst/millennium-seed-bank.

Royal Botanic Garden Kew. n.d. [a]. 'Millennium Seed Bank.' Accessed 1 August 2019. www.kew.org/science/collections-and-resources/research-facilities/millennium-seed-bank.

Shaffer, H. 2018. 'Urban biodiversity arks.' *Nature Sustainability* 1: 725–7.

Sjöman, H., Hirons, A.D., Bassuk, N.L. 2015. 'Urban forest resilience through tree selection — Variation in drought tolerance in *Acer*.' *Urban Forestry & Urban Greening* 14 (4): 858–65.

Song, J., Murdoch, J., Gardiner, S.E., Young, A., Jameson, P.E., and Clemens, J. 2008. 'Molecular markers and a sequence deletion in intron 2 of the putative partial homologue of *LEAFY* reveal geographical structure to genetic diversity in the acutely threatened legume genus *Clianthus*.' *Biological Conservation* 141 (8): 2041–53.

Statistics New Zealand. 2017. 'Agricultural production statistics by farm type, June 2017.' Accessed 27 February 2020. www.stats.govt.nz/assets/Uploads/Agricultural-production-statistics-/Agricultural-production-statistics-June-2017-final-/Download-data/agricultural-production-statistics-jun17-final-additional-tables-2-farm-counts.xlsx.

Thompson, C.W. 2011. 'Linking landscape and health: the recurring theme.' *Landscape and Urban Planning* 99 (3–4): 187–95.

Timm, S., Dearborn, L., and Pomeroy, J. 2018. 'Nature and the city: measuring the attention restoration benefits of Singapore's urban vertical greenery.' *Technology|Architecture + Design* 2 (2): 240–9. https://doi.org/10.1080/24751448.2018.1497377.

Ulrich, R.S. 1999. 'Effects of gardens on health outcomes: theory and research.' In *Healing Gardens,* edited by Clare Cooper Marcus and Marni Barnes, 27–86. New York: John Wiley and Sons.

van de Walt, K. 2018. 'The *ex situ* conservation of recalcitrant Myrtaceae species, a response to myrtle rust incursion in the Pacific region. Unpublished PhD confirmation report. Palmerston North: Massey University.

Westpac Institutional Bank. 2016. 'Industry insights: horticulture.' 28 July 2016. Accessed 29 June 2018. www.westpac.co.nz/assets/Business/Economic-Updates/2016/Bulletins-2016/Industry-Insights-Horticulture-July-2016.pdf.

Widrlechner, M.P. 1994. 'Environmental analogs in the search for stress-tolerant landscape plants.' *Journal of Arboriculture* 20 (2): 114–119.

Widrlechner, M.P. 2004. 'Insights into woody plant adaptation and practical applications.' North Central Regional Plant Introduction Station conference papers, posters and presentations. Iowa State Digital Repository. Accessed 27 February 2010. https://lib.dr.iastate.edu/ncrpis_conf.

Wood, E., Harsant, A., Dallimer, M., Cronin de Chavez, A., McEachan, R.R.C., and Hassall, C. 2018. 'Not all green space is created equal: biodiversity predicts psychological restorative benefits from urban green space. *Frontiers in Psychology* 9: 2320. https://dx.doi.org/10.3389%2Ffpsyg.2018.02320.

Chapter 10
Weeds

Kerry Harrington

Chapter 10
Weeds

Kerry Harrington

School of Agriculture and Environment, Massey University

Introduction

In many agricultural and horticultural situations, such as pasture, arable crops, vegetable crops, orchards, gardens and lawns, weeds can cause problems and need to be kept under control. Weeds can be defined as plants growing where they are not wanted, thus interfering with human activities. Whether or not a plant species is considered a weed can depend on the situation. A species such as white clover is highly desirable in pastures where it increases the quality of fodder for livestock and also fixes atmospheric nitrogen using rhizobia (bacteria) located in nodules on its roots. But white clover isn't wanted in flower gardens, where it looks messy, nor in lawns where it can attract bees to its flowers which may result in bee-stings for people walking on the lawn with bare feet.

In this chapter we shall explore the reasons why weeds aren't wanted in agriculture and horticulture, study a few aspects of their biology, and then look briefly at how weeds can be controlled.

Problems caused by weeds

A major effect of weeds is that they compete with desirable plants for light, soil nutrients and water (Figure 10.1). When desirable plants are just establishing from seed, weeds can often establish more rapidly, growing taller than the pasture or crop plants and thus shading them from light. As the plants get bigger, the amount of nitrogen, phosphate, potassium and other nu-

Figure 10.1 A tomato plant (middle right) struggles to establish among weed seedlings, mainly nettle and redroot.

trients in the soil required for their growth may start to become limiting. Many weed species can absorb these nutrients quite rapidly, depleting the amounts left for crop and pasture plants. As it becomes drier in summer, weeds can also use up the water present in the soil, resulting in water stress in crop and pasture plants occurring sooner than if weeds were absent. Some weed species exude allelopathic substances, which are chemicals that can make other plants grow less aggressively. Competition is generally much more important than allelopathy, but it is an additional effect of weeds in some situations. The combined effects of competition and allelopathy can result in greatly reduced crop yields if weeds are not controlled.

Competition is only one problem caused by weeds, however. A weedy crop will often also have more problems with pests and diseases. Weeds growing around the edge of paddocks can act as host plants for pests and diseases that need somewhere to live during winter, when crop plants are not growing or (if deciduous species) have dropped their leaves. Once the growing season gets under way, the pests and diseases move back into the crop. A weedy crop has less air movement near the ground, which means that the relative humidity within that weedy crop increases, and a high humidity encourages fungal growth. Thus, a disease outbreak will be more likely in a weedy crop (Figure 10.2). Although predators such as birds or parasitic insects attack insect pests, when a crop is weedy then the predators and parasites cannot find the insect pests as easily.

Most crops are machine-harvested, and the

presence of significant numbers of weeds generally decreases harvest efficiency. There are several factors. First, grain needs to decrease in moisture content before being harvested. As mentioned above, the level of humidity is higher in weedy crops, which can decrease the rate of drying and thus delay harvesting. Second, if large amounts of weed material are pulled into the harvester along with the crop, separation of the grain or fruit from other material becomes more difficult. Third, some weeds, such as scrambling fumitory (*Fumaria muralis*) or cleavers (*Galium aparine*), clamber up over the top of a crop, making it top-heavy so it may fall over in the wind. This makes it more difficult for the harvester to pick up grain, as it is now closer to the ground. Fourth, hand-harvesting occurs with some vegetable and fruit crops and pickers will find the work unpleasant if prickly thistles and stinging nettles are present. Weeds can also cover the fruits, not only making them harder to find but also potentially stopping fruit crops developing their correct colour through poor exposure to light.

Other quality issues can occur during harvesting. For example, the berries of black nightshade (*Solanum nigrum*) might get harvested along with peas destined for processing. These berries can be difficult to separate from the peas, and if they ended up in packets of frozen peas then

Figure 10.2 The dense black nightshade growing between these sweet corn plants increases humidity, making disease outbreaks more likely.

consumers would worry that their health might be affected. Many members of the public believe that these berries come from deadly nightshade (*Atropa bella-donna*), which is actually rare in New Zealand. Although a few black nightshade berries should not cause much harm, food companies do not want contaminants in their produce.

Many crops are harvested for seeds to be sown by other farmers. Weed seeds harvested along with the crop seeds need to be separated out if possible to avoid reducing the quality of the crop seed. Some weed seeds are very similar in shape and size to crop seeds, making separation difficult and costly. The presence of the seeds of some weed species in crop seed will prevent it from being exported.

In pastures and grazed crops such as lucerne the presence of some weed species will reduce the amount of grass, clover and crop species available to be eaten. Also, animals avoid eating plants close to prickly weeds like thistles or near poisonous weeds like ragwort. High densities of thistles mean that quite a bit of pasture or crop is wasted as the animals won't eat it (Figure 10.3).

Weeds can also decrease the quality of products from grazing animals. Dairy cows eating weeds such as twin cress (*Lepidium didymum*) will produce milk with an unpleasant taint. Prickly weeds can get caught up in wool and reduce its quality. Some weeds, such as barley grass (*Critesion* spp.), have seed-heads that work through the wool of sheep into the skin of the animals, causing festering sores that reduce the quality of both the meat and the leather produced from the skin. Although animals usually avoid poisonous weeds, sometimes they do eat them, causing health problems.

In perennial fruit crops, weed problems

Figure 10.3 Thistle species such as Scotch thistle (above right) and Californian thistle (foreground) can make it difficult for animals to graze pasture underneath them.

Figure 10.4 This boysenberry crop would be more productive if it wasn't swamped by redroot plants.

include competition for nutrients and water, hosting of pests and diseases by weeds through winter, decreased harvesting efficiency, and sometimes poorer fruit quality if weeds prevent fruits being exposed to light (Figure 10.4). Another issue is that many of these crops can be sensitive to frost when new growth appears in early spring. Bare soil below trees and vines is heated during the day by the sun, and this heat is released at night and may reduce frost damage. Weeds covering the soil prevent this heating taking place which makes frost damage more likely.

Having a lot of weeds on a farm gives the impression of poor management. Many farmers control weeds because they enjoy a weed-free working environment, and also do not want neighbours thinking that they are incapable of keeping weeds under control.

Weeds are not just a problem in agriculture and horticulture. In pine forests, weed competition can cause newly planted pines to be severely suppressed, increasing the number of years before timber can be harvested. Spiny weeds like gorse (*Ulex europaeus*) and blackberry (*Rubus fruticosus*) make operations like thinning and pruning of trees more difficult,

and forest fires are more likely with weeds drying out in summer and allowing fires to spread. Weeds make gardens and lawns look less aesthetically pleasing, and some weed species in lawns can put prickles in bare feet. Weeds can cause visibility problems along roadsides and may trip people up on footpaths. Species such as old man's beard (*Clematis vitalba*) and tradescantia (*Tradescantia fluminensis*) can degrade the quality of native bush. Weeds in waste places make them look untidy, cause fire hazards, can harbour pests such as rats, and can produce pollens that cause hay fever. Aquatic weeds degrade streams, rivers and lakes, making them less pleasant for swimming and fishing, and can make drains work less efficiently (Figure 10.5).

Figure 10.5 Water moves less efficiently through drains that are clogged with weeds.

Characteristics of weeds

Although any plant species could be classified as a weed if it were growing where it is not wanted, some plant species typically cause weed problems much more commonly than other species. Weed species will generally be different in pastures, which are frequently grazed, and in crops that are planted into bare soil each year and harvested a few months later.

Plant species are annuals, biennials or perennials. An *annual* species completes its life-cycle after about 6 months and then dies once seed has been produced. Some annuals may complete their life-cycle even more quickly, whereas others may last almost a year before they finally die off. Most of our problematic cropping weeds are annual species (e.g. fathen *Chenopodium album*, redroot *Amaranthus powellii* and black nightshade). Because they need to establish each year from seeds in the soil, it suits their life-cycle to establish in the bare soil prepared for sowing crops.

Many weeds are summer annuals; they usually germinate in spring, then flower in summer before producing seeds and dying in autumn. They are usually unable to tolerate frosts. Black nightshade is a summer annual which usually establishes in mid-spring. Occasionally plants of black nightshade germinate in late summer — these need to complete their life-cycle very rapidly because of the day-length starting to decrease, so can produce seeds within about 6–8 weeks. However, if black nightshade grows under trees or in other areas protected from frosts, they can sometimes get through a winter before dying and so last nearly a year.

Winter annuals mainly establish in autumn, produce seed in early spring and die before dry summers commence. These species often do not tolerate dryness well (e.g. barley grass in pastures).

Other annual species, such as scrambling speedwell (*Veronica persica*), can germinate at any time of the year because New Zealand doesn't get as cold or as dry as the parts of the world where these species originate.

Biennials are species that take longer than one year to complete their life-cycle. They flower and set seed in their second year of growth, then die off following seed production like annual weeds do. In agriculture, biennial weeds include ragwort (*Jacobaea vulgaris*), Scotch thistle (*Cirsium vulgare*) and nodding thistle (*Carduus nutans*).

Perennial weed species survive longer than two years. Some perennial species have lignin in their cells, making them woody, so they have permanent trunks and branches above ground. Woody perennial weeds that cause problems in agriculture include scrub species such as gorse, broom and blackberry (*Rubus fruticosus*).

Other perennial species do not have woody stems, and tend to replace their leaves and stems at regular intervals. Some of these have quite permanent below-ground structures, such as taproots (e.g. docks *Rumex* spp. and dandelion *Taraxacum officinale*), while others have creeping underground stems called rhizomes (e.g. couch *Elytrigia repens*; Figure 10.6). Other perennial weeds have underground creeping root structures (e.g. Californian thistle *Cirsium arvense*), or bulbs (e.g. some species of oxalis), a few have tubers (e.g. purple nutsedge *Cyperus rotundus*), and some have stolons, which are creeping stems that grow along the top of the soil (e.g. creeping buttercup *Ranunculus repens*). Quite a few perennial weeds grow in clumps with

Figure 10.6 Couch is a difficult weed to kill because of the rhizomes (underground stems) that easily regrow should plants be disturbed.

Figure 10.7 High densities of scrambling speedwell have established in this kale crop from dormant seeds in the soil, and further seeds are about to be produced and put back into the soil.

fibrous root systems (e.g. Yorkshire fog *Holcus lanatus*). Apart from the taproots, most of these other structures tend to replace themselves every year or so, resulting in plants that can be many years old but with no parts that are more than a few years old. This is known as vegetative reproduction.

Perennial weeds usually start out as seedlings, but once established they are difficult to control as they generally have a lot of stored carbohy-

drate in structures such as their roots, rhizomes and bulbs which allows them to regrow aggressively if they get disturbed. They tolerate life in pastures and lawns very well because they don't have to establish from seed each year. Annuals struggle to re-establish from seed regularly in pasture due to competition from the pasture sward. Perennials tend to establish quite slowly from seed compared with annuals, so are not found in crops as much as annual weeds are. However, some perennial weeds, such as couch, docks and oxalis, are well adapted to regrow from their underground systems after seedbed preparation for cropping.

Of the thousands of plant species in the world, those that typically cause weed problems grow competitively and aggressively. They are adaptable, often being able to grow well under a wide range of growing conditions. They thrive in disturbed habitats, being able to regrow quickly either from seeds in the soil or from underground structures such as rhizomes, bulbs and taproots. They have efficient reproduction, usually producing many thousands of seeds per plant which last many years in the soil, and often also with good underground vegetative reproduction structures in the case of perennial weeds. Successful weed species are also good at surviving unfavourable conditions.

Most New Zealand weeds come from overseas countries where it can be very cold in winter, or very dry at some times of the year, or from a country where agriculture has been undertaken for many centuries. Plant species that can survive and thrive in such environments will also do well in agriculture where there is a lot of cultivation and herbicide spraying, and in pastures where they are exposed to intensive grazing. Their prolific production of seeds that can last many years in the soil (Figure 10.7), and extensive vegetative reproduction with structures that can remain dormant underground for many months also helps them to thrive.

Weed control

So, how are weeds controlled in agriculture and horticulture? Spraying with herbicides is the main weed-control technique, but there are a few other options. These should be used in conjunction with herbicides for more efficient weed control, and can sometimes be used instead of herbicides.

Preventative weed control (biosecurity)

As mentioned above, most of the troublesome weed species in New Zealand have come from overseas countries. Some arrived here accidentally, often along with European settlers during the late 1800s. Others were intentionally introduced, such as gorse for hedges, blackberry as a fruit crop and sweet briar (*Rosa rubiginosa*) as an ornamental garden plant. There is now much better control over material coming into New Zealand, with strict quarantine systems to stop people bringing in weeds or their seeds either accidentally or intentionally. As New Zealand is isolated from the rest of the world by many kilometres of sea, it is hopefully possible to prevent other weed species arriving here, as there are many troublesome species not yet in New Zealand. However, failures occasionally occur with the quarantine system and new species do still arrive.

Only a fraction of all the weed species within New Zealand are present on each farm and horticultural unit in the country. Every effort should be taken to prevent new weeds from arriving. Seeds for some weed species can travel in the wind, and others have berries which get spread by birds. But generally, species move only short distances in this way.

Weed seeds usually arrive in a new area within crop seed, so farmers should buy certified seed that has been tested for the presence of weed seeds. New species can also arrive in hay; farmers should buy hay locally, where weed species are likely to be similar. Seeds of new weed species can also travel on harvesting equipment and cultivation equipment, so these should be thoroughly washed before entering a farm. If a new plant species growing on a property is noticed, it should be controlled promptly so that the plants don't produce seeds and start building up a population.

Discouraging weed growth through management techniques

Weed species that have already arrived and are commonly found on a property won't cause problems each year until the seeds within the soil germinate and establish. Most weed seeds have dormancy mechanisms to stop them germinating at times when there is little chance of them establishing successfully. These mechanisms have evolved to help the species become successful; if they are understood, they can be used to manage weeds.

One common mechanism is that many seeds will not germinate if the light above the seed has been filtered through green vegetation. This filtering increases the ratio of far-red light compared with red light. Under these conditions, any seeds that germinated would find themselves in the shade of established plants. The seedlings would then be likely to suffer from severe competition and die within days of germination. Staying dormant until the ratio of red light increases gives the seedlings better conditions for growth.

Farmers try to keep their pastures dense throughout the year so that the soil is shaded by the pasture plants, thus preventing germination of weeds. Weeds often become established when the pasture is damaged in some way, such as drought stress in summer (allowing germination when it rains in autumn) or treading damage by animals in winter. Farmers also use a number of techniques to get their crops established as rapidly as possible so that the weed seedlings are more likely to be out-competed. Early formation of a dense crop canopy also stops further weed seeds from germinating. More rapid establishment of crops can be achieved by ensuring good preparation of seedbeds, not sowing seeds too deeply, ensuring that sufficient fertiliser is applied and sowing at the optimal time of year with respect to temperature and rainfall.

Many horticultural crops need to be well spaced to give good crop yields, so out-competing weeds or stopping seeds from germinating through forming a dense canopy is not an option. Another dormancy mechanism found in many seeds is not germinating if buried too deeply. Most weed seeds are small, so do not have sufficient food reserves to grow up to the light from too deep in the soil. With some crops it is feasible to cover the soil with mulch material, such as sawdust, straw or black polythene, to block light from reaching the soil and thus keep seeds dormant (Figure 10.8). Even if they did break dormancy, seedlings would struggle to penetrate up through these layers.

Figure 10.8 A combination of polythene mulches within crop rows and straw between rows helps keep this strawberry crop free of weeds.

Once weeds have germinated and established, management techniques can help reduce their dominance, especially in pastures where competition can be exerted by the pasture sward and grazing can be used to discourage weeds. The optimal conditions for plant growth can differ between species. For example, the ryegrass and clover growing in pastures do not regrow rapidly after grazing if the soil is waterlogged, whereas a number of weed species, such as buttercups and docks, tolerate waterlogged soils very well and so can dominate pastures under these conditions. Putting in a drainage system removes the competitive advantage for those weeds that tolerate wet conditions, increasing the chances of the pasture out-competing the weeds. Likewise, if soils become too acidic, or too low in phosphate or nitrogen, there are weed species that tolerate these conditions better than pasture. Lime should be applied to deal with acidity, and fertiliser can help with nutrient deficiencies.

Perennial ryegrass (*Lolium perenne*) and white clover (*Trifolium repens*) are two of the best plant species for tolerating both grazing and treading damage. By using the correct grazing intensity and frequency, many weeds can be stopped from dominating because they do not handle grazing as well as the ryegrass and clover. However, over-grazing of pastures allows weeds such as thistles to become established in bare soil, while under-grazing leads to scrub weeds such as gorse becoming a problem because they aren't removed by grazing during their vulnerable seedling stage. Animal species vary in their preferences. Cattle don't like eating ragwort or buttercups, but sheep will graze these; goats will eat woody weeds that neither sheep nor cattle will eat. Thus, using different classes of stock can help control weeds.

Biological control of weeds

Weed species arriving in New Zealand from overseas have usually left behind many of their natural enemies such as insects and diseases, so often grow better in New Zealand than in those other countries. Two forms of biological control can potentially be used to combat these weeds.

Classical biological control involves finding suitable organisms, such as insects or fungi, in the country a weed originates from and then introducing them to New Zealand to attack the weed. Only organisms that feed exclusively on a particular weed species are permitted to enter New Zealand, as otherwise there is a risk that the organisms will cause damage to beneficial or native plants. As a result, classical biological control usually fails to cause much damage to weeds — organisms that have evolved to feed on only one species are not likely to destroy all plants of that species, as otherwise they would not survive very long themselves (Figure 10.9). Occasionally, weed species do become less prolific as a result of these organisms, as has occurred with St John's wort (*Hypericum perforatum*),

Figure 10.9 Gorse continues to be a troublesome weed in New Zealand forestry despite the release of eight different biological control species in recent decades in an attempt to control it.

a poisonous weed in pastures.

Inundative biological control attempts to multiply organisms (usually fungi) that already exist in New Zealand up to levels that will be more damaging to weeds than the natural equilibrium levels. These organisms are sprayed on to a paddock to cause a disease outbreak. Despite considerable research, however, no examples have currently been successful for weed control in New Zealand as it is difficult both to store the organisms and to apply them to reliably give high levels of control.

Physical weed control

Weeds can also be controlled by physical force. The main technique is through the use of cultivation implements, and these have been used for weed control since biblical times.

The simplest device is the hand hoe, used by home gardeners and council workers tending flower gardens to cut the stems of weed plants from their root systems and to uproot young weeds. They are occasionally used in high-value horticultural crops where herbicides or tractor-driven implements can't be used, but labour is generally too expensive in New Zealand to justify the use of hand hoes in crops. They are used extensively in overseas countries where labour is cheap and other forms of weed control are too expensive for some farmers to afford.

Most weed control in commercial agriculture and horticulture within New Zealand makes use of mechanised equipment, usually mounted on the back of a tractor. A common implement for killing weeds in a paddock before establishing a crop is the mouldboard plough. This implement slices through the soil and tips the slices upside down in furrows, so that plants that were grow-

Figure 10.10 A mouldboard plough can help control weeds by burying them.

ing on the surface get buried (Figure 10.10). Other implements use power from the engine of the tractor to rotate blades in a trailed implement such as a power harrow or a rotary hoe; these blades cut up weeds, flick some weeds out of the ground and bury others.

Ploughs and rotary hoes usually control weeds across a complete paddock to prepare the ground for establishing a crop. Other equipment can be used to cultivate the soil between rows of crop plants. These implements, known as inter-row cultivators, vary in complexity from a simple blade pulled along to slice weeds, thus separating shoots from roots, through to modified rotary hoes with rotating blades designed to kill weeds between crop rows.

Most cultivation implements will kill annual weeds successfully, especially if they are only seedlings. Perennial weeds are much more difficult to kill, however, as they can often regrow from vegetative propagules such as rhizomes and taproots. Multiple passes of cultivation equipment are required to exhaust the root reserves of these weeds, so translocated herbicides such as glyphosate (see the next section) are usually favoured for the control of perennial weeds.

Mowing is another form of physical weed control, often done to control weeds growing between rows of fruit trees and vines. It is also sometimes used in pastures to help manage upright flowering thistle stems and rushes. Mowing seldom kills weeds, but it can stop them becoming too dominant and will usually reduce their production of seed.

Some thistle species can be controlled using a chipping hoe to cut 2–3 cm below the dormant buds located within the crowns found at ground level, which is usually enough to kill the weed. Plants such as flowering ragwort can be pulled out of the ground to prevent seed production, but both pulling and chipping are labour-intensive exercises.

Chemical weed control

Generally, the most effective and efficient form of weed control is the application of herbicides.

Types of herbicide

Some herbicides are *non-selective* (e.g. glyphosate) — these kill all plants to which they are applied. In contrast, *selective* herbicides can be applied over both crop and weeds, and these kill just the weeds, not the crop (e.g. chlorsulfuron in wheat). Non-selective herbicides can be applied selectively, however, with a good example being glyphosate which is the main herbicide used in apple orchards and vineyards. If glyphosate was applied to the foliage of apples or grapes, the crops would die. But instead it is only applied to the weeds below the trees and vines, taking care not to get any spray on the leaves of the crops.

Some herbicides are *foliar-applied* — they are applied to the foliage of existing weeds, from where they penetrate into the leaves and kill the weeds. Other herbicides are *soil-applied*; these are usually applied to bare soil after a seedbed has been prepared and the crop has been sown, but before the crop or weeds emerge. Most soil-applied herbicides are also known as *pre-emergence* herbicides; they work by killing germinating seedlings but are often poor at controlling existing weeds. Although the term 'pre-emergence' can refer to application before emergence of the crop, in perennial fruit crops such as apples and grapes the term refers only to the emergence of weeds (as the crops have usually been present for many years). Foliar-applied herbicides are often also known as *post-emergence* herbicides.

For herbicides that are applied to the foliage of weeds, some move very little within the plant but simply scorch those parts of the plant to which they are applied. These are known as *contact* herbicides. An example of a non-selective contact herbicide is paraquat, while bentazone is a selective contact herbicide. Contact herbicides don't travel into the roots of weeds, so generally are only good for controlling annual weeds (perennial weeds can regrow from their roots). Weeds are most susceptible to contact herbicides when they are seedlings.

The other type of foliar-applied herbicide is the *translocated* herbicide, also sometimes called a *systemic* herbicide. After penetrating into leaves, these herbicides will move within the plant to other parts, and often the aim is to get them to move into the root system to give good weed control. Translocated herbicides can vary in their mobility within plants. Glyphosate is an example of a translocated herbicide that translocates very well, compared with 2,4-D which translocates reasonably well but usually not well enough to kill most well-established perennial weeds. These herbicides typically move with the sugars being transported around the plant, so work best when plants are growing actively — at this time lots of sugars are being produced and moved around the plant (Figure 10.11).

Some herbicides are deactivated as soon as they reach soil, with glyphosate and paraquat being the best examples of this. Most other herbicides leave residues in the soil for a while after application; this is particularly important in the case of soil-applied herbicides as otherwise they would not remain active for sufficiently long to kill the weed seedlings as they germinated.

Many factors can influence how long herbicide residues remain active in the soil. Some herbicide types are more persistent than others; the persistent ones are not broken down easily. Most, however, are broken down within a few weeks or months by microorganisms in the soil (such as bacteria and fungi), and this occurs most quickly in warm, moist conditions.

Herbicide safety

The general public perception is that herbicides are extremely dangerous to use, which is unfortunate because this is not the case. The situation has not been helped by the recent public debate in the media regarding whether or not glyphosate causes cancer.

Before any pesticide (including herbicides, insecticides and fungicides) can be sold in New Zealand, a regulatory body known as the Environmental Protection Authority (EPA) must assess all available data on the risks of using that pesticide within the country. The EPA will then determine whether the chemical can be sold here, and whether there needs to be restrictions over who can buy and use the chemical. Only some of the safer pesticides can be bought from outlets such as garden centres and supermarkets and used by all members of the public. The rest can only be used by farmers and growers who have undergone a training course, for which they receive a certificate showing that they are competent to use the chemicals. The EPA will usually ban chemicals that have been shown to cause cancer or birth defects, or those that are extremely toxic.

The safety of chemicals is determined by conducting tests over several years involving a range of different animal species. These tests reveal how toxic a chemical is, and whether, among other things, it causes cancer or eye damage, forms residues in foods we eat or persists in waterways. The results of the animal testing are used to predict the likely effect on humans who might be exposed to the chemical through applying it, eating plants that have been sprayed with it, or being exposed to spray drift.

Most of our commonly used herbicides are less toxic than aspirin or common salt, and none cause cancer. The debate over glyphosate started in 2015 when a group of toxicologists published a report claiming that glyphosate may cause cancer in humans. At least two other reports have since been published by other groups of toxicologists, in which they say that the initial report was flawed and there is no evidence of glyphosate causing cancer. Most of the animal testing has given no indication that we should be concerned. Even if the original group of toxi-

cologists were correct, according to their scoring system glyphosate is still less likely to cause cancer than drinking alcohol, eating bacon or spending time in sunshine.

However, it remains important to read the label on all herbicides, as many require users to wear gloves to protect against rashes, and some require eye protection when measuring the concentrate as splashes can cause eye damage. Most importantly, users should take care not to allow spray droplets to drift, as they can damage nearby crop plants, and herbicides should generally be kept away from streams and other waterways.

How herbicides are used

To illustrate how herbicides can be used to control weeds, a few typical weed-control programmes will be briefly described.

To establish pastures based on perennial

Figure 10.11 The creeping root system of Californian thistle makes it very difficult to control, so translocated herbicides are needed to kill the roots.

ryegrass and white clover, glyphosate is used to remove existing vegetation as it kills a wide range of species and leaves no residues in the soil. The ground may then be cultivated, or ryegrass and clover seed may be direct-drilled into the sprayed ground. Once the ryegrass and clover have established, they may be dense enough to out-compete weeds. However, usually there are sufficient weeds present to justify applying a selective post-emergence herbicide, with MCPB being most commonly used as it kills the seedlings of many weeds such as thistles without causing damage to the young grass or clover. If high densities of weeds are present in later years, a selective herbicide such as 2,4-D may be applied; this kills older weeds more effectively than MCPB, but can cause some temporary suppression of clover. Herbicides that are even more effective than 2,4-D are available, but these kill clover outright (e.g. clopyralid). Accordingly, these are sprayed only on to the weeds, known as *spot-spraying*.

In some arable crops, such as wheat, a selective post-emergence herbicide may be applied across the field if weed densities are high once the crop has established. Many herbicides can be used for this, so a mixture that controls most of the weeds present is often selected. Pre-emergence herbicides are available for some arable crops, such as maize; a mixture of atrazine and acetochlor is traditionally used to kill weeds as they emerge, and post-emergence herbicides such as nicosulfuron may also be applied later if some weeds still manage to establish.

Grasses and weeds are allowed to grow between the rows of fruit crops such as apples and grapes, and are then mown at regular intervals to reduce competition with crop roots growing under this area. Having bare soil between the crop rows for many years can lead to degradation of the soil, and organic matter from the mown plants will prevent this degradation. The ground near the base of the trees or vines is usually kept bare of vegetation, however, to reduce competition and to make it easier to mow between the rows. Glyphosate is the main herbicide used to kill weeds within the crop rows. As some weeds can tolerate glyphosate, other herbicides can be alternated with the glyphosate. Amitrole, another non-selective translocated herbicide, or glufosinate, a non-selective contact herbicide, are often used. A soil-applied herbicide such as terbuthylazine may also be applied to kill new weeds as they germinate.

Problems and future issues

Although this chapter has shown that weeds can be controlled within New Zealand using a variety of methods, the control of weeds is often not straightforward. The high cost of labour within New Zealand means that many of the more labour-intensive strategies for weed control cannot really be justified, so there is heavy reliance on herbicides to control weeds efficiently. There is general public misunderstanding about the relative safety of herbicides along with mistrust of the companies that produce them. As a result, there is constant pressure globally for herbicides to be used less frequently or preferably not at all. If herbicides such as glyphosate were to be banned, however, greater use of alternative weed control techniques such as cultivation would be required, and these are generally more expensive and less effective.

Persistent use of some herbicides has led to a build-up of herbicide resistance in some populations of weeds (Figure 10.12). Although not as bad in New Zealand as in some other countries, herbicide resistance cases are being reported more frequently here. As the development of new herbicides is so expensive, very few new products are making it on to the market. Accordingly, ongoing research is investigating other ways to control weeds, as well as how to make the most efficient use of the techniques currently available and ways to stop further herbicide resistance from developing.

Figure 10.12 Continuous use of glyphosate for control of vegetation under grapevines is now resulting in weeds developing resistance to this herbicide in parts of New Zealand.

Further reading

More information on many of the weed species mentioned in this chapter, including what they look like and how to control them, can be found at www.massey.ac.nz/weeds.

Champion, P., James, T., Popay, I., and Ford, K. 2012. *An illustrated guide to common grasses, sedges and rushes of New Zealand.* New Zealand Plant Protection Society.

Popay, I., Champion, P., and James, T. 2010. *An illustrated guide to common weeds of New Zealand* (3rd ed.). New Zealand Plant Protection Society.

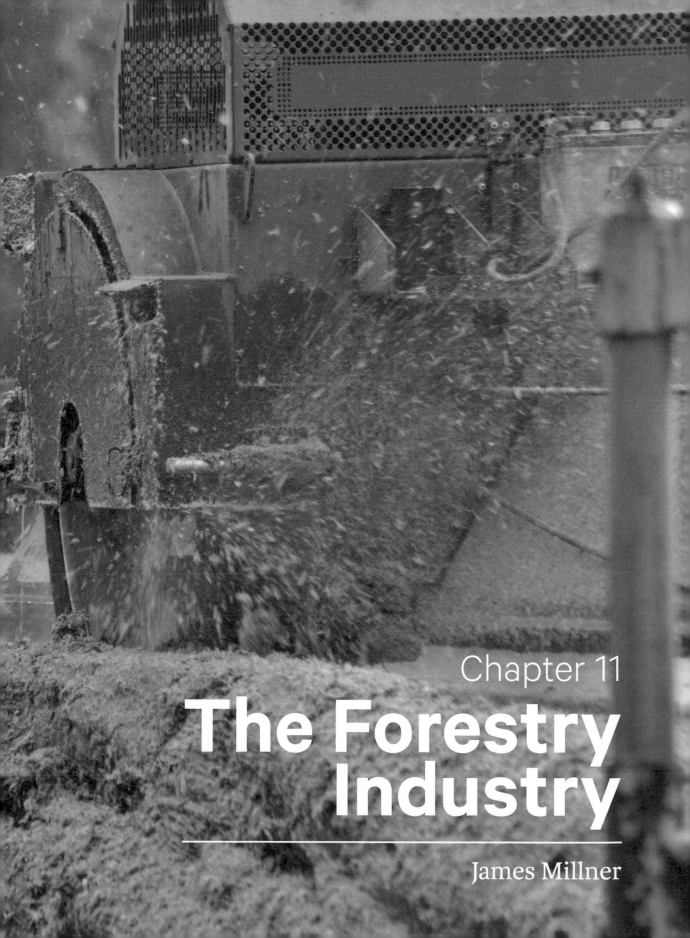

Chapter 11

The Forestry Industry

James Millner

Chapter 11
The Forestry Industry

James Millner

School of Agriculture and Environment, Massey University

Introduction

Most of the agricultural land in New Zealand was originally covered in rainforest which was almost completely removed during the conversion to agriculture (see Chapter 2). New Zealand's forestry industry was closely associated with this activity; mostly through the selection and extraction of commercially valuable logs, after which the cutover bush was burned before being oversown with pasture grasses and legumes. The main species extracted were podocarps with useful wood characteristics, e.g. rimu (*Dacrydium cupressinum*), mataī (*Prumnopitys taxifolia*), kahikatea (*Dacrycarpus dacrydioides*) and tōtara (*Podocarpus totara*). In addition, kauri (*Agathis australis*), a conifer, was heavily cut in Northland and Coromandel. South Island beech forests (see Figure 11.1) were mostly exploited in the 1900s for woodchips, which were exported.

A large number of broadleaf species also occur in New Zealand forests, with the most abundant tall species being tawa (*Beilschmiedia tawa*), kāmahi (*Weinmannia racemosa*) and rātā (*Metrosideros* spp.).

In the late 1800s it became clear that forest resources were becoming exhausted; most of the remaining unlogged native forest was in remote areas with difficult terrain. The need to establish plantations (planted forests) to ensure adequate supplies of wood and timber in the future was identified. These plantations involved exotic tree species, rather than native species, because native species have very slow growth. Most native trees would not reach commercial size until well over 100 years of growth (Table 11.1), compared with less than 30 years for the better-performing exotic species. *Pinus radiata* (radiata

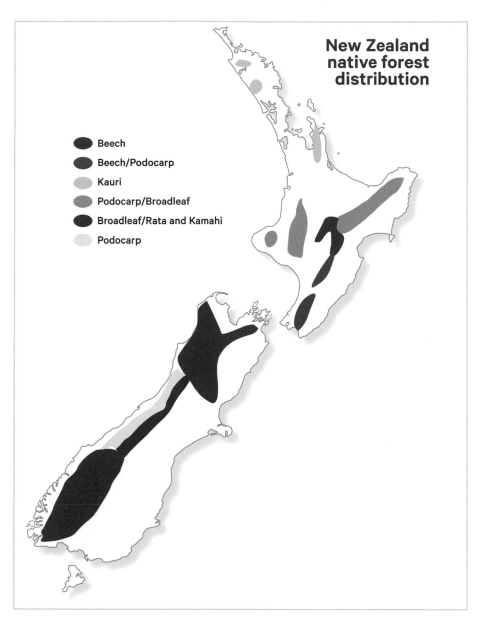

New Zealand native forest distribution

- ● Beech
- ● Beech/Podocarp
- ● Kauri
- ● Podocarp/Broadleaf
- ● Broadleaf/Rata and Kamahi
- ● Podocarp

Figure 11.1 Current distribution of New Zealand native forests.

Table 11.1 Size of two native tree species under-planted in a degraded *Pinus ponderosa* plantation after 50 years growth in the central North Island compared with typical radiata pine.

	Rimu	Tōtara	Radiata pine
Height (m)	11.5	10.6	40
Stem diameter (cm)	20.1	15.0	50

(Forbes, Norton, and Carswell 2014)

pine) in particular quickly gained a reputation as a species with very fast growth, suited to a wide range of sites and able to thrive in low-fertility soils. However, it is unsuited to areas with high snowfall because of snow damage to the canopy.

Radiata pine is a very versatile species utilised for a wide range of end uses, from paper to furniture. An important disadvantage of radiata pine is that it has very low natural durability or resistance against decay, especially if wet or in contact with the ground. This vulnerability to decay has been a problem in the construction industry, where a combination of poor workmanship and house designs that are vulnerable to leaking has resulted in many houses throughout New Zealand not being weather-tight (known as leaky buildings). In these houses, any framing timber not adequately treated to prevent decay has deteriorated and requires substantial rebuilding to correct. If used outdoors or in situations where there is a risk of it getting wet, radiata pine must be treated with a chemical preservative. The most popular preservative in New Zealand is chromated copper arsenate (CCA).

Plantation forests

Today, almost all of New Zealand's wood production comes from plantation forests. This is unique globally because other countries with sizeable forest industries, for example Russia and Canada, still rely heavily on natural forests which are usually allowed to regrow after harvest. Some countries are not utilising their forests sustainably, resulting in ongoing deforestation. These are mostly developing countries in tropical regions, for example Brazil and Indonesia. Deforestation is a major contributor to increasing atmospheric CO_2 levels. About 50 per cent of the dry weight of wood is carbon.

Radiata pine accounts for about 90 per cent of our national plantation area, mainly because its rapid growth allows trees to be harvested at less than 30 years of age. Foresters in many other parts of the world often wait over 100 years between planting and harvesting if producing logs for milling.

Initially the plantation forest area in New Zealand increased slowly, but the early 1930s saw a boom in planting as a result of using unemployed people to plant trees as an employment scheme during the Great Depression. During this time plantations were confined to land regarded as having no value for pastoral farming, particularly on the Volcanic Plateau north of Taupō.

Although plantation forests are now spread throughout New Zealand, the bulk of the forest area remains in the North Island. Currently the plantation area is estimated at 1.71 million ha and is expected to increase as new plantations are established. Most plantations and woodlots are replanted after harvest, but there has been some conversion of forests to dairy on the Volcanic Plateau and in Canterbury where land is of gentle topography. Land on the Volcanic Plateau was not suitable for livestock production originally because of cobalt deficiency, while in Canterbury very dry, stony soils and lack of irrigation encouraged forestry. However, cobalt deficiency is now easily corrected and irrigation is now widely available in Canterbury.

Forestry expansion is mostly occurring on marginal hill country. For highly erosion-prone hill country this means a halt to the degradation that has plagued this class of land ever since the removal of the original native forest. Plantation forests are able to reduce slip erosion by up to 90 per cent in most regions. Additionally, plantation forests are able to sequester large amounts of carbon, helping to reduce atmospheric CO_2 levels and potentially generating income from the sale of carbon credits.

Plantations may consist of different stands of trees of different species, different management regimes and different ages. A good example of this can be seen in Kaingaroa Forest between Rotorua and Taupō. Within each stand, usually an area of land bounded by roads and known as a *compartment*, the trees will be the same species and age, managed similarly and harvested at the same time. A key objective of plantation management is to have as little variation within a compartment as possible.

Tree species

Radiata pine

Radiata pine is an exotic conifer or softwood species originating from California and is the most commonly planted timber species in New Zealand. With good management, it is capable of producing marketable logs in 25 to 30 years on most New Zealand sites; this is faster than any other commercial species. The duration between planting and harvesting is known as the *rotation length*.

The wood is pale in colour with moderate strength but has poor natural resistance to decay. The dominance of radiata pine is the result of its high productivity and the versatility of the wood. The important attributes of radiata include the following:

- Seedlings are easy and inexpensive to produce
- High growth rates
- High seedling survival rates after planting (>90% is typical)
- Tolerant of a wide range of environments
- Wide range of end uses, including:
  - sawn timber
  - fibreboard
  - plywood
  - paper and cardboard
  - posts, poles and sleepers
  - laminated veneer lumber (LVL)
  - furniture.

Douglas fir

Douglas fir (*Pseudotsuga menziesii*) is the second-most important forestry species in New Zealand but only occupies 6 per cent of the national plantation area. It is an important species in high-country sites, particularly in the South Island, because it is able to cope with low temperatures and high snowfall. Whereas radiata pine catches snow, which builds up so that eventually the tree collapses under the weight, Douglas fir is able to shed snow. Most Douglas fir is utilised for framing timber. It is regarded as one of the world's best species for this end use because it produces relatively stiff timber which resists bending under pressure. It is slower-growing than radiata pine, with rotation lengths often being 50 to 60 years to produce marketable logs.

Others

A large number of species have been planted or trialled, but areas planted in individual species are small so make very little contribution to the supply of logs. Most of these so-called alternative species are grown on farms and include cypresses (e.g. macrocarpa *Cupressus macrocar-*

Table 11.2 Contribution of different forest products to New Zealand's total forestry industry export revenue (December 2019).

Forestry products	Percentage of total industry value
Panels	7.4
Logs and wood chips	56.0
Wood pulp	12.0
Paper and paperboard	7.0
Sawn timber and sleepers	13.6
Other forestry products	4.0

Source: New Zealand Forest Owners Association

pa and Mexican cypress *Cupressus lusitanica*), eucalypts (e.g. brown barrel *Eucalyptus fastigata* and yellow stringybark *Eucalyptus muelleriana*) and coastal or California redwood (*Sequoia sempervirens*). While many of these species are more difficult to grow than radiata pine, being slower-growing and more vulnerable to pests and diseases, they may have superior wood quality for specific end uses.

Economic importance

Plantation forests occupy about 7 per cent of New Zealand's land area, making forestry a significant national industry. By global standards the New Zealand forestry industry is small, contributing about 2 per cent of traded forest products globally. Most forest products produced in New Zealand are exported (85 per cent); in 2018 the value of exports was about $6.4 billion, with exports to China in particular having increased greatly in recent years. A range of products are exported (Table 11.2). Log exports have grown markedly, whereas processed forest products (e.g. paper),

which generate greater income and employment in New Zealand, have declined. This is an issue for the industry. China takes mostly unprocessed logs, whereas Australia mostly takes sawn framing timber used in house construction.

The domestic market is important for some forest products; for example, most sawn timber and plywood produced is used within New Zealand. There is also an important domestic market for 'round wood', which is used for posts and poles.

Plantation management in New Zealand

The forestry industry in New Zealand has been dominated by large private companies, often with overseas owners. However, recent expansion has mainly occurred through investment by individual farmers and groups of investors in woodlots and small plantations. Farmers collectively own a significant share of the plantation estate (180,000 ha in 2018), albeit in small, fragmented woodlots. The lack of size is a disadvantage, as the cost of accessing and harvesting small, remote forest blocks on steep country sometimes means that they are not economic to harvest. The establishment of a carbon market as part of the New Zealand Emissions Trading Scheme can help improve the economic viability of forestry in remote areas. Some land-owners have established plantations solely for the purpose of generating income through the sale of carbon credits rather than logs.

Plantations are managed to produce the logs required by both the domestic sawmilling industry and the key overseas log markets. Logs are

Clockwise from top left: radiata pine, Douglas fir, macrocarpa and coastal redwood.

Table 11.3 Export log grades for radiata pine based on pruning, minimum small end diameter (SED) and knot diameter.

Grade	Minimum SED (cm)	Maximum knot diameter (cm)	Length (m)
Pruned			
P1	40	-	4–6
P2	30	-	4–6
Unpruned			
A	30	12	6–8
K1	26	25	6–8

defined by whether or not they contain clear-wood (wood free of knots), log diameter, and diameter of knots if present (Table 11.3). Other important log traits include straightness. Douglas fir is very rarely pruned because currently there is no market for clearwood logs.

When selling logs, the diameter of the smallest end of the log is the measurement that most influences the value. Tree stems usually taper so that stem diameter is greatest at ground level and least at the tip. That means that logs have a large end, closest to the ground, and a small end at the upper end of the log. The small end is more important than the large end because it determines how much of the log can be used for cutting boards of the same length as the log.

Management regimes

New Zealand's *Pinus radiata* plantations are planted with genetically improved seedlings and intensively managed with weed control and tending regimes designed to produce marketable logs for specific end uses. Tending operations may include thinning and pruning, and are usually completed by age 8–10 years in most plantations. Clear-felling or logging occurs from about 25 to 28 years after planting and after harvest most plantations are replanted.

Plantations may be managed to produce a range of different logs, including pruned logs, unpruned logs and pulp logs.

- High-quality *pruned* logs (used for furniture and joinery) are produced from plantations that have been pruned and thinned to produce butt logs (the bottom log) without knots (clearwood regime). However, a typical tree from a clearwood regime will produce a range of logs of different sizes and grades depending on the management of the tree during the first 7–8 years of growth (Figure 11.3). Note that pruned and unpruned saw-logs are produced as well as logs that are only suitable for pulp.
- Framing regimes produce *unpruned* sawlogs only; these are milled to produce timber suitable for house framing, packaging and boxing. Trees grown under this regime are thinned but not pruned.
- Plantations may also be managed to produce *pulp* logs — yield is the key consideration and there is no thinning or pruning.

Most plantations in New Zealand have traditionally been managed using a clearwood

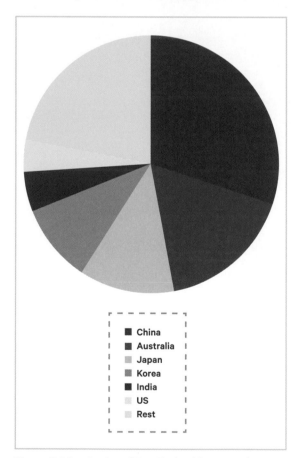

China
Australia
Japan
Korea
India
US
Rest

Figure 11.2 Destination of New Zealand forest products based on value (2017). (Adapted from New Zealand Forest Owners Association, Facts and Figures 2017/18)

regime, but declining premiums for pruned compared with unpruned logs has resulted in more stands not being pruned or pruned to a lower height, e.g. 4.5 m rather than 6.5 m. The value of different log grades can be found on the website of the Ministry for Primary Industries (see the end of the chapter for details).

Pruned logs have the greatest value per cubic metre and vary in price according to the small end diameter (SED). Unpruned logs differ in value according to SED and the maximum diameter of the branches or knots in the log. There are additional costs involved with producing pruned logs (e.g. pruning), and the current market returns for these logs make clearwood (pruned) regimes arguably still the most profitable in most situations. It is uncertain whether this market advantage will continue; strong house-construction activity in New Zealand and overseas helps because this requires a lot of clearwood for products such as doors, window frames and furniture.

A typical clearwood programme for a fertile site is outlined in Table 11.4. Establishing a plantation begins with planting. Usually around 1000 seedlings/ha will be planted, but this may be greater on sites where seedling losses might be high, such as drought-prone areas. This planting rate allows managers to select the best trees to become the final crop trees, based on growth and form during thinning operations. Most seedlings planted in New Zealand are of the GF (Growth and Form) breed. Seedlings are rated from GF 1 (unimproved) to GF 30 and higher. High-GF seedlings are the result of planned crosses between parents with superior growth and form. Weeds are controlled soon after planting, usually using a selective herbicide applied as a spot spray over the seedlings.

The objective of clearwood regimes is to

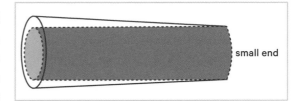

Figure 11.3 The importance of the small end diameter of a log with respect to sawing. The wood outside of the central cylinder is lost when boards the full length of the log are milled.

Table 11.4 Example of a clearwood tending programme for a farm site.

Age (years)	Height (m)	Operation
0	–	Plant 1000 stems/ha (4 m row spacing with 2.5 m between seedlings in each row)
		Weed control (spot spray)
4	6	Prune 400 stems/ha to 2.2 m
		Thin to 600 stems/ha
6	9	Prune 350 stems/ha to 4.2 m
8	12	Prune 350 stems/ha to 6.0 m
		Thin to 350 stems/ha
28	35	Harvest

confine knots associated with branch growth to a defect core at the centre of the tree, usually less than 20 cm in diameter (depending on site), while maximising the growth of valuable clearwood over the defect core by timely thinning (Figure 11.4). Pruning must be supported by thinning to reduce competition in the stand, to allow trees selected for good form and growth to increase in diameter and produce clearwood. Competition reduces diameter growth rather than height growth in trees.

Tending regimes vary with the site index (see the section on site productivity, below), basal area

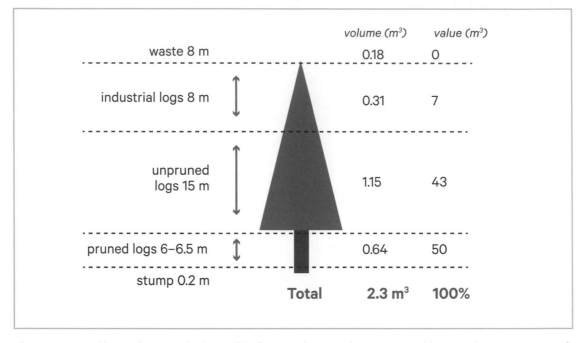

	volume (m³)	value (m³)
waste 8 m	0.18	0
industrial logs 8 m	0.31	7
unpruned logs 15 m	1.15	43
pruned logs 6–6.5 m	0.64	50
stump 0.2 m		
Total	**2.3 m³**	**100%**

Figure 11.4 Typical log production and value profile of a pruned *Pinus radiata* tree. Pruned log contributes 50 per cent of the value of the logs in the tree despite lower volume (0.64 m³) than the unpruned saw logs. (Adapted from New Zealand Forest Owners Association, Facts and Figures 2017/18)

potential (likewise, see below), topography, weed populations, the genetic quality of seedlings and the market expectations for different log grades.

Tending operations

The scheduling of tending operations such as thinning and pruning is a critical part of the management regime. If pruning is done too late, the proportion of clearwood in the log may be too low and it will be downgraded to unpruned. Thinning is crucial for reducing competition between trees, to encourage diameter growth, so late thinning results in lower diameter at harvest, lower yields at harvest or delayed harvest (increased rotation length).

Thinning is mostly 'thinning to waste', which results in culled trees being left to decay after being cut. 'Production thinning' may be carried out in some stands when harvesting costs are low (gentle topography and close to markets) and prices for round wood are high.

Pruning is usually scheduled according to a target stem diameter at the lowest unpruned whorl of branches. On a sample of trees, the diameter over the pruned branch stubs of this whorl is measured and compared with a target diameter. Once the target is reached, then pruning occurs on a stand. This target diameter is known as the diameter over the (pruned) stubs or DOS. The DOS determines the size of the defect core at the centre of a pruned log and may be around 18–20 cm for a fertile site (most farms) but less than 15 cm for a less-fertile site (e.g. sand country). The defect core is the zone at the centre of a log that contains the pruned branch stubs and any zone of distorted wood grain resulting from occlusion of branch stubs (Figure 11.5). Pruning needs to be a balance between minimising the size of the defect core, minimising the number of pruning lifts required, and maximising diameter growth.

Trees are typically pruned to between 6 and 6.5 m in two to three pruning lifts. Above this height the generation of clearwood after pruning is insufficient to cover the cost of pruning, which becomes increasingly expensive as pruned height increases. The target DOS should be the same for each lift.

On more-fertile sites it is often difficult to restrict the number of lifts to just two or three. Because the tree grows in diameter more quickly on such sites, the target DOS is reached at a younger age and consequently the tree is not as tall, reducing the height of each pruning lift. Pruning live branches from a tree reduces diameter growth for up to 2 years after pruning, but over-pruning a tree (leaving too few live branches) will severely reduce tree growth (diameter) so that the pruned tree will not be able to compete with neighbouring unpruned trees. Most pruning lifts aim to leave 3–3.5 m of green canopy on the tree to ensure adequate photosynthesis.

Some forest owners are moving away from pruning because the value of the clearwood produced as a result of pruning is not currently covering the cost of pruning, in terms of both labour and reduced log volumes.

Harvesting

Harvesting or logging is a crucial activity in forestry; the rewards of 25–30 years of investment and management are finally realised. It requires careful planning and a high degree of opera-

Figure 11.5 Cross-section of a pruned log showing the extent of the defect core (indicted by the arrow) surrounded by clearwood. Clearwood is used for higher-value end uses (e.g. furniture), so is more valuable than knotty wood.

tional skill because of the financial and physical risks involved. Logging begins with trees being cut down, which is increasingly being done by machines even on very steep hill country. Different logging systems are based on how logs are transported back to the site where they are cut up and loaded onto trucks (the skid site) and delivered to either a New Zealand sawmill or a port for export. There are two broad logging categories: ground harvesting or cable harvesting.

Ground harvesting is usually done with the aid of machines called skidders. These are able to move around quickly, dragging logs from the stump to the skid site where they are cut up according to specifications. Logs are often produced by machines that are able to remove branches, measure the key log traits such as length and diameter, and cut them up to maximise their value. This system is the least expensive way to harvest logs in most situations but is not suited to hill-country sites, especially steep hill country, where cable harvesting is used.

Cable harvesting systems rely on a large cable

Cable harvesting using a grapple hook.

strung from a high point, such as a ridge, over the slope to be harvested. Once trees are cut, logs are attached to the cable using strops or grapple hooks and winched up to the skid site. Logs are always extracted uphill because there is minimal risk of being hit by falling logs. Cable harvesting is slower and therefore less productive than harvesting with the use of skidders, and also requires more-expensive machinery (e.g. winches). Consequently, it is more expensive than ground-based systems but is often the only option in much of New Zealand's steep hill country. Another disadvantage of traditional cable harvesting systems is that people are required to work under the cables on steep slopes, attaching logs to the cable with strops prior to winching; the risk of death and injury from falling logs is significant.

The use of grapple hooks fitted with remote sensing technology to extract logs in hill country is an example of the measures being adopted to increase productivity and reduce deaths and injuries in forestry — these don't require people on the ground to attach the logs to the cable. However, the logs do have to be cut with a feller-buncher, which grasps each tree, cuts it (feller), and then places it on the ground (buncher) to be recovered by the grapple. These machines can work in very steep country especially if they are tethered to the top of the slope with a wire cable.

Feller-buncher operating in hill country with bunched logs in the foreground.

Site productivity

Potential tree growth and wood yield is determined by a number of management and environmental factors. The important management factors include species choice, genetic potential of seedlings and stocking rate (trees/ha). Production also increases with age of trees up until the stage where growth in wood volume is matched by losses due to tree deaths, which may not occur until stands are over 50 years of age. Most plantations are harvested well before this stage.

The environmental factors influencing tree growth fall into two categories: those affecting height growth, indicated by the *site index*, and those influencing diameter growth, indicated by *basal area potential*.

Site index

Site index has a major influence on potential forest productivity. In New Zealand the site index is nearly always given as the height (in metres) of the largest-diameter trees in a *Pinus radiata* stand at 20 years of age. It is generally independent of management with the exception of very low stocking rates, which tend to reduce tree height. Height growth has a large influence on wood volume, with sites having a high site index generally producing greater volumes of wood at any given age and stocking rate. In other countries, a site index may refer to the growth of different species over a much longer time span, particularly in regions where tree growth is slow.

Site index is determined by environmental factors including temperature, rainfall and soil fertility. In New Zealand the site index can vary from <25 (poor) to >30 (good) in commercial forests. Low values are mostly found in the southern South Island, while high-index sites are found in warm, high-rainfall regions in the North Island.

Basal area potential

Basal area potential describes the potential for diameter growth and is influenced by soil fertility and stand age, and management inputs such as stocking rate (trees/ha) and species. Fertile sites produce larger-diameter logs than low-fertility sites when comparing the same species at the same age and stocking rate. Basal area potential is expressed as the total cross-sectional area of all trees in 1 ha at a height of 1.4 m and is expressed as m²/ha. It can be estimated by measuring the mean diameter and basal area of trees in a stand and multiplying by the stocking rate (trees/ha).

The diameter of a tree trunk at 1.4 m is known as diameter at breast height (DBH). It is usually found by measuring the circumference of the trunk at that height and converting to diameter by dividing by pi (3.14). This approach assumes that the trunk is approximately round.

High basal areas tend to be a characteristic of farm sites, primarily because of the high fertility status of soils that have been previously used for pastoral farming. Log values, particularly for unpruned logs, may be low on these sites because of stem malformation, heavy branching and wind damage. There is a strong relationship between log diameter and branch size — big logs mean big branches.

Basal area potential can be used to estimate stocking rate. A high-fertility site may be capable of supporting a stand basal area of 70 m²/ha, whereas a lower-fertility site may only support a stand basal area of 50 m²/ha at the intended age of harvest. This information allows foresters to estimate the number of trees per hectare that can be carried if the stands are to produce trees with similar diameters at harvest. If the stocking rate is too high then diameter growth will be low, so the trees will take longer to reach the required DBH. This will delay harvest, which is undesirable because it usually reduces profitability. To produce logs with small end diameters of >30 cm (see Table 11.3), the DBH needs to be around 50 cm. The stocking rate (trees/ha) which can be carried on different sites to achieve this mean DBH can be estimated using the calculations outlined below.

Stand basal area	=	70 m²/ha
Mean DBH	=	50 cm or 0.5 m
Mean basal area	=	0.20 m²/tree
Stocking rate (70 m²/ha ÷ 0.20 m²/tree)		
	=	350 trees/ha

Mean basal area is calculated as $(\pi \times DBH^2)/4$, where DBH is in metres. For a site with a basal area potential of 50 m²/ha, the stocking rate would be 250 trees/ha.

Health and safety

Forestry is one of the most dangerous industries for workers. It often involves working in steep, relatively inaccessible country, using large machinery to undertake work that carries a lot of risk, especially logging. The number of workers killed or injured is a significant ongoing issue for the industry (Table 11.5). One of the common causes of deaths in forestry is trees falling on workers during logging operations. Despite the introduction of safer workplace practices and the requirement for workers to undergo industry training, a significant number of people are killed or seriously injured every year.

The response to the dangers in the forestry industry have ranged from increased emphasis on health and safety training of workers, through increased scrutiny of logging crews from Worksafe inspectors to the adoption of new technologies, especially for harvesting, to minimise the number of people involved in dangerous operations such as felling. The use of feller-bunchers for cutting trees is an example of this.

Environmental issues

Wilding conifers

Many of the important plantation species, including radiata pine and Douglas fir, are capable of producing seed and spreading naturally once established. They become weeds. This is now a large problem in much of the high country in both the North and South Islands. For pastoral farmers, the spread of wilding conifers means a reduction in grazing area; for the Department of Conservation, it means loss of habitat for rare or endangered native plants and animals.

The total area affected by wilding conifers in New Zealand is estimated to be about 1.8 million ha. Control efforts to date have included aerial application of herbicides and cutting, but the scale of the infestation and the inability of land-owners, including the Department of Conservation, to fund control operations has meant that wilding pines are continuing to spread. In the future it may be necessary to use conifer breeds that don't produce seed in regions where wilding trees are likely to become weeds.

Carbon sequestration

Trees store, or sequester, carbon — about 50 per cent of the dry biomass of wood is carbon. The source of this carbon is atmospheric CO_2 which is fixed by the canopy of trees (photosynthesis) to produce biomass. In trees and other woody plants, much of this biomass is converted to wood. The ability of trees to remove and store atmospheric CO_2 makes them potentially very useful in the global quest to limit atmospheric CO_2 levels to help limit global warming. CO_2 is the most significant greenhouse gas in the atmosphere. In New Zealand, carbon emissions and

Table 11.5 Number of people killed or seriously injured (requiring more than 1 week away from work) in the New Zealand forestry industry from 2011 to 2017.

	2011	2012	2013	2014	2015	2016	2017
Injuries	207	165	177	153	147	108	129
Deaths	3	5	7	4	3	4	7

Source: Worksafe.govt.nz

sequestration are expressed as CO_2 equivalents. One tonne (t) of CO_2 is classified as 1 emissions unit. One tonne of carbon = 3.6t CO_2.

Plantations have been seen as a way of storing atmospheric CO_2 and enabling New Zealand to meet its commitments under the Paris Agreement, which aims to limit global temperature increases this century to 1.5–2°C above pre-industrial levels. New Zealand agreed to reduce CO_2 emissions to 30 per cent below its 2005 levels, a challenging target. This target could be achieved by reducing emissions as well as by sequestering atmospheric CO_2. The rules governing the treatment of emissions and the sequestration of CO_2 are set out in the New Zealand Emissions Trading Scheme.

Achieving the target by reduction in emissions alone would be very difficult and costly. Young, rapidly growing trees are capable of sequestering large amounts of carbon. For example, a 28-year-old radiata pine plantation in the North Island may hold the equivalent of 700–800 t/ha of CO_2. Plantation owners can register the increase in stored carbon (carbon credits) and sell them to industries that emit CO_2. Prices have ranged from >$20/tCO_2$ to <$5/tCO_2$ in recent years. When prices are high there is incentive for land-owners and investors to establish forest plantations to sequester carbon. Most of these forests are on steep, erosion-prone hill country, resulting in dual benefits: reduced erosion plus income from the sale of carbon credits and logs. This can generate significant income while the trees are growing, a big advantage in an industry with very long payback periods, but a sizeable proportion of sold carbon credits must be bought back when the forest is harvested. This is because sequestered CO_2 is eventually released back into the atmosphere after harvest, and owners are required to account for this loss.

If the plantation is replanted, the impact is reduced because the new crop immediately starts to sequester carbon again. An outline of the role of forestry in the New Zealand Emissions Trading Scheme can be found at the link provided at the end of the chapter.

With the recent renewed emphasis on meeting New Zealand's international commitments to reduce CO_2 emissions, the value of New Zealand carbon credits has increased to about $27/t CO_2, sparking renewed interest in forestry for sequestering carbon.

Biosecurity issues

A number of pests and diseases not presently in New Zealand have the potential to do huge damage to the national forestry industry. There have been incursions in the past, mostly from pests and diseases that are transported from Australia to New Zealand on the prevailing westerly winds. Recent examples include the arrival of the fungal disease myrtle rust, which poses a threat to many native species belonging to the myrtle family, including mānuka (*Leptospermum scoparium*) and pōhutukawa (*Metrosideros excelsa*), as well as introduced eucalypt trees. Many insect pests of eucalypts have established in New Zealand and some have become economically serious, for example the brown lace lerp (*Cardiaspina fiscella*) which can greatly reduce vigour in *Eucalyptus saligna* and *Eucalyptus botryoides*, two useful timber species. Our most important species, radiata pine and Douglas fir, are also vulnerable to the arrival of new pests and diseases; threats include pitch pine canker (*Fusarium circinatum*), which in North America can kill over 50 per cent of trees in infected forests. Important insect pests with the potential to cause severe damage to plantation and native forests include Asian gypsy moth (*Lymantria dispar*). This moth has been found in New Zealand, but was detected sufficiently early to permit elimination.

Plantation mānuka

Most people think of forestry as being about the production of logs, but forests produce many other direct and indirect non-timber outputs. Examples of indirect outputs include reduced sedimentation of streams and rivers after rain due to the ability of trees to stabilise slopes, and increased biodiversity. Direct benefits might include food production. Some New Zealand land-owners are establishing mānuka to produce honey, rather than timber species.

Mānuka has traditionally been regarded as a weed; many hill-country farmers will have had to control it at some time. Conservationists have always promoted its role in native forest restoration. Mānuka is a pioneer species, being among the first species to appear after a major disturbance such as fire or erosion, so is adapted to harsh sites. It acts as a nurse crop for forest species, provides good habitat for many native bird species and offers good protection from erosion in hill country.

Over the past 10–15 years, mānuka has become an important crop for many apiarists because of the high prices being achieved for mānuka honey, particularly honey with a high Unique Mānuka Factor (UMF) content. This is based on the proven health benefits of high-UMF mānuka honey. The UMF value arises from the presence of methylglyoxal, which has strong antiseptic properties and is only found in mānuka honey and some honey from Australia, also produced from *Leptospermum* species. The health benefits are primarily associated with the ability of high-UMF honey to fight persistent skin infections and aid healing of wounds.

The demand for high-value mānuka honey is driving the establishment of mānuka plantations; thousands of hectares have been planted over the past few years. For hill-country farmers this may generate significant income from honey or hive rental, carbon credits and protection of highly erodible soils. See the end of the chapter for a link to useful information about this rapidly growing industry.

Research

Forestry research is primarily undertaken by Scion, a Crown Research Institute based in Rotorua. Canterbury University also makes a significant contribution to forest research, and Massey and Lincoln Universities are also involved. Research spans a wide range of topics that include plantation management, genetic improvement, control of pests and diseases, wood properties and environmental benefits (e.g. carbon sequestration).

The future

The forest industry will continue to grow in the short term and the area planted will increase, partly because of initiatives such as the One

Billion Trees Programme which is an important strategy in New Zealand's effort to reduce net carbon emissions. New technologies that allow wood to be used for multi-storey buildings will help lift demand for wood in the construction industry. In the longer term, increased investment will be needed so that more logs can be processed domestically to reduce reliance on changeable export markets for unprocessed logs.

References

Forbes, Adam S., Norton, David A., and Carswell, Fiona E. 2014. 'Underplanting degraded exotic *Pinus* with indigenous conifers assists forest restoration.' *Ecological Management & Restoration* 16 (1): 41–9.

Worksafe Mahi Haumaru Aotearoa. www.worksafe.govt.nz.

Further information

- Forestry facts and figures: www.nzfoa. org.nz/resources/publications/facts-and-figures.
- Forest management: www.nzffa.org.nz/ farm-forestry-model/the-essentials.
- Forest species: www.nzffa.org.nz/farm-forestry-model/species.
- Log prices: www.teururakau.govt.nz/ news-and-resources/open-data-and-forecasting/forestry/wood-product-markets.
- Plantation mānuka for honey: www. manukafarmingnz.co.nz.
- Role of forestry in the Emissions Trading Scheme: www.teururakau.govt.nz/ funding-and-programmes/forestry/emissions-trading-scheme.
- Wilding conifers: www.doc.govt.nz/nature/ pests-and-threats/weeds/common-weeds/ wilding-conifers.

Precision Agriculture

Pullanagari Rajasheker Reddy

Chapter 12

Precision Agriculture

Pullanagari Rajasheker Reddy

Massey AgriFood Digital Lab, School of Food and Advanced Technology
Massey University

Introduction

Providing food for the world's rapidly growing population continues to be challenging due to increased pressure on limited land and water resources as well as scientific constraints. Conventional farming relies heavily on intensive, unsustainable use of water and fossil-fuel-based chemicals, resulting in major environmental costs. We need sustainable agricultural systems that can continue to provide growth in the supply of food within defined ecological, economic and social limits.

Precision agriculture (PA) is a potential alternative method to boost agriculture production through using advanced smart/hi-tech technologies (Gebbers and Adamchuk 2010). These technologies will improve management strategies through better understanding of plant and environment interactions and the spatial variability on both a paddock and a farm scale (Shannon, Clay, and Kitchen 2020).

Precision agriculture will increase farm productivity and offer a risk solution to optimise farm profitability and reduce environmental damage. Precision agriculture uses sensors, the global positioning system (GPS), geographic information system (GIS), auto-steering tractors, smart robots, yield monitors, microprocessors, computers, cameras and more.

Farm productivity is typically affected by variability. This is a measure of dissimilarity in plant-, soil- and environment-related characteristics and their complex interactions. Typical plant characteristics are yield, pest/disease resistance and vigour. Soil characteristics are organic matter, nutrients, texture, moisture, etc. Environmental characteristics include topography, rainfall and temperature. Farmers need knowledge of these characteristics when making effective decisions to manage variability. However, the characteristics are highly variable across the farm in both spatial and time domains. The key goal of PA is to identify the spatial and temporal variability and then tailor the appropriate management to that variability.

Precision agriculture in cropping follows the 4Rs approach: supplying the *Right* source of fertiliser in the *Right* place at the *Right* application rate and at the *Right* time (Shannon, Clay, and Kitchen, 2020).

Soil nutrients are typically supplied through chemical fertilisers such as urea, single superphosphate (SSP), di-ammonium phosphate (DAP), lime and others. In conventional agriculture, a single composite soil/plant sample collected from the field is the basis for fertiliser application, with fertiliser being applied uniformly over the field. Through ignoring the variability, however, this process of uniform fertiliser application applies overdoses of fertilisers in areas of high fertility and lower-than-required doses in areas of low fertility.

Precision agriculture is primarily focused on cropping, however, it can relate to any agriculture production system, including horticulture, livestock farming, dairy farming, hill-country farming, viticulture, aquaculture and forestry.

Global navigation satellite system (GNSS)

A GNSS is the most commonly used technology in PA, where it is used to identify the position of agricultural and environmental operations.

GNSS is a satellite-based technology that transmits radio signals between satellites and receivers. These receivers are mounted on farm vehicles or can be carried by hand. The signals are transmitted at the speed of light, 299,792 km/second (Yousefi and Razdari 2015). Based on the time the signals take to travel from the satellite to the ground receiver, the location of the object (e.g. a tractor) on the Earth's surface is calculated.

Geolocation is the estimation of the geographical location of an object based on coordinates, latitude and longitude (Figure 12.1). On the Earth's surface, each point/object has unique coordinates referred to by numbers, letters or symbols.

- Latitude lines circle the Earth parallel to the equator, which runs east to west. Latitude ranges from −90 at the North Pole to +90 at the South Pole.
- Longitude lines run north to south lengthways down the Earth's surface and converge at the poles. Longitude ranges from −180 to +180.

A geographic coordinates system is a reference system that enables the location of any object on the Earth's surface using a mathematical model based on latitude and longitude. For example, the geographical coordinates of Wellington city are −41.2865° S, 174.7762° E. If we include elevation with the latitude and longitude information, the position is referred to as a 3D location.

Various GNSS receivers are available to suit different agricultural practices, from soil sampling to autonomous farm vehicles, and provide different levels of position accuracy depending on what is required for the activity (Table 12.1). Despite its higher cost, the RTK-GPS type has been widely adopted by many farmers as they can see the direct benefits from the higher specifications of these devices.

Several different satellite-based positioning systems have been developed globally by different countries: GPS by the United States of America, GLONASS by Russia and GALILEO by Europe. The technology was initially created for military applications, being adopted in agriculture in the early 1990s. In New Zealand, GPS technology is now widely used for seed drilling, pasture measurement, fertiliser and pesticide spraying, animal tracking and crop harvesting.

Geographic information system (GIS)

A geographic information system (GIS) is a computer system for capturing, storing, analysing, checking and displaying data related to positions on the Earth's surface. In PA, GIS helps famers to visualise and analyse agricultural data (yield, plant vigour, nutrient availability) with reference to the geolocation (latitude and longitude)

Table 12.1 GNSS systems available in the market for agricultural applications.

Type	Accuracy	Cost	PA applications
Standalone GPS	6–10 m	Cheap	Soil sampling, farm location, fencing
DGPS	1–5 ms	Moderate	Soil sampling, fencing, yield mapping, fertiliser application
RTK-GPS	0–5 cm	Expensive	Land levelling, auto-steering, vehicle guidance, ploughing, sowing

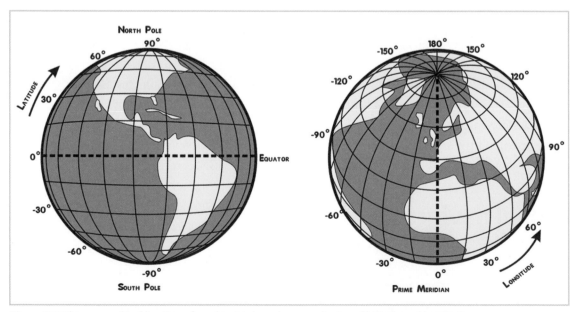

Figure 12.1 The geographical location of an object is based on coordinates of latitude and longitude. (Source: Wikimedia Commons/djexplo)

(Yousefi and Razdari 2015). GIS tools are powerful tools that can help farmers to understand spatial variability throughout the farm, and its causes, by analysing the data — thus enabling effective management decisions to be made.

In PA, two types of data format are widely used in GIS: vector data and raster data. Vector data formats are used to create digital features such as points, lines and polygons to represent objects. For example, farm boundaries are represented by polygonal features indicating area, while soil-sample or tree locations are represented by point features (no area or length). Farm tracks, streams and drainage channels are line features, indicating location and length but not area. These objects are geo-referenced using GPS tools, which calculate the coordinates and therefore the position of the object. Geo-referencing is a key concept that underlies all GIS functions.

GIS has thousands of tools that can be used to analyse data from different objects. Measurements are made at specific points within a field, each point being linked with a geographical location. Yield maps are a simple example of the use of point data (Figure 12.2). To provide better understanding of the data, the point measurements are converted into a smoothed continuous map using GIS software. In this process, data anomalies are also removed to represent field variability.

Raster data formats are grids of cells or pixels producing a series of rows and columns (Figure 12.3). In raster data, such as elevation or satellite imagery, each pixel represents a specific part of the Earth's surface with a unique coordinate. Pixel size varies based on sensor configuration. Raster images are usually created using a camera/sensor.

Decision-making is an important step in pre-

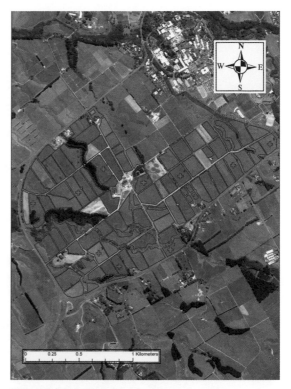

Figure 12.2 Point data for a farm.

Figure 12.3 Raster data for a farm.

cision farming that encompasses data collection, data analysis and interpretation. Generally, data are collected with different tools and the data locations are geo-referenced using GPS/GNSS. GIS stores all the collected data and displays it in

Figure 12.4 Thematic map of the pasture dry matter percentage of a Massey University dairy farm.

different layers for conducting further analysis. Data analysis summarises and organises the data into actionable information as reference and thematic maps.

- Reference maps encompass information about boundaries, fencing, roads, streams, farmhouses and aerial photographs.
- Thematic maps display field variability information in different themes/colours as categories. A simple thematic map of pasture dry matter percentage is shown in Figure 12.4; each category is assigned a different colour. The resolution of thematic maps depends on the scales used for the data layers. These digital maps are flexible and colours can be adjusted using a GIS software.

Yield monitors

Yield monitors are sensors mounted on harvesting machines to monitor and estimate the flow of a commodity (e.g. grain, herbage, cotton lint, fruits of different crops). Yield monitors are combined with GPS receivers to obtain total yield per location. A yield monitor comprises sensors, a converter to translate the sensor information into yield data, storage, a display and a GNSS/GPS receiver (Figure 12.5). These sensors generate yield information that can be used to create yield maps using GIS software (Figure 12.6). Every data field has a certain amount of variability represented within such yield maps. When the yield maps are loaded into GIS software, users can trace linkages to yield-limiting factors such as topography, moisture and nutrient deficiency, or pest or disease infestation.

Soil variability and fertility management

Soil variability is principally caused by several factors, including inherent differences between soil types, the results of management approaches, variable topography and uneven application of fertiliser. To understand soil variability and apply fertiliser most efficiently, soil sampling is typically conducted (Whelan and Taylor 2013).

Sampling strategies

For precision management applications, in-field sampling is used to collect soil nutrient information. Compared with traditional agriculture, PA requires intensive sampling for detailed information. This is carried out using a grid sampling approach, where the farm is divided into small grids overlaying homogeneous areas. Grid size is dependent on user requirements and the scale of management decisions. A grid sampling strategy uses sufficient dense samples to reveal fertility levels and patterns within the field. To represent each grid appropriately, composite samples are recommended rather than single cores.

Although grid sampling is an effective approach for depicting soil variability on a small to large scale, performing high-density sampling is expensive and time-consuming. To reduce the number of samples, the alternative approach of zone sampling can be used. In zone sampling, a field is divided into different zones and a limited number of samples is collected from each zone.

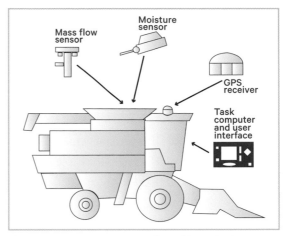

Figure 12.5 Basic components of a yield-monitoring system (Adapted from Grisso, Alley and McClellan, 2005).

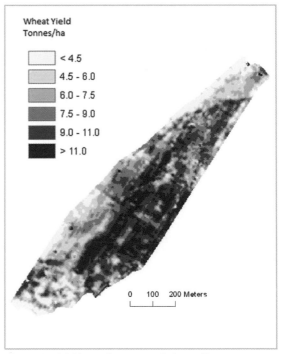

Figure 12.6 Yield map for a crop of wheat. (Source: LandWISE, New Zealand)

Several tools are available to delineate the management zones, including topographic maps, satellite imagery, drone imagery, yield maps and sensor information. The zones become management units; each zone is treated as an unbroken area to which specific management treatments (e.g. fertiliser, fungicide, irrigation) are applied overall. Within each zone the variability is less than the variability of the whole field, and is therefore not considered significant in terms of whole farm variability management. When creating management zones, multiple sources of information are used for precise and reliable outputs, rather than relying on a single source of information. Advanced statistical tools are also used to find the patterns (spatial and temporal) in different layers and consolidate these patterns into meaningful management zones.

Topographical data

Topography is an important component that influences crop productivity and the availability of moisture and nutrients. For example, rainwater runs downhill from ridges, hilltops and slopes and accumulates in depressions. Along with rainwater, silt and some nutrients also accumulate in those areas. As a result, plant growth is at a maximum within depressions compared with hilly regions of the field. Typically, topography is defined by satellite radar or airborne LiDAR (light detection and ranging) data.

Satellite imagery is widely available to describe field variability. Earth observation satellites regularly collect surface information at varying resolutions (1–1000 m). For example, the Landsat satellite images the entire Earth every 16 days in an 8-day offset with a resolution of 30 m. Landsat has been collecting these images since 1972, and since then hundreds of satellites have been launched by different agencies. In recent years, the use of unmanned aerial vehicles (UAVs) or drones has grown rapidly, enabling the capture of high-resolution images cheaply and whenever required.

Combined multi-year yield maps have been used as a tool to delineate zones. These historical yield maps also help farmers to highlight areas where additional management is required.

Tools in precision agriculture

In PA, remote sensing tools are widely used to assess crop health or variability. Remote sensing means collecting information about some property of an object using a device/sensor that is not in physical contact with the object. Researchers have proposed various types of platforms for operating such devices (Figure 12.8):

- Ground platforms, also known as proximal sensing
  - Operate a few metres away from the objects
  - Stationary — sensors are used as handheld devices or fixed in a location
  - On-the-go — sensors can be moved across the field when mounted on a vehicle, such as a tractor, all-terrain vehicle or sprayer
- Airborne platforms
  - Unmanned aerial vehicles (UAVs or drones) — operate only a few metres above the Earth's surface
  - Fixed-wing aircraft — operate hundreds to thousands of metres above the Earth's surface
- Satellite platforms
  - Operate from orbit hundreds of kilometres above the Earth's surface,

e.g. Landsat satellites launched by NASA.

Each platform has specific advantages and disadvantages. Satellite-based sensors can cover large areas in a short time with medium to coarse resolution. UAVs and handheld/proximal sensors can provide high-quality information with high resolution. Moreover, UAVs and proximal tools have operational flexibility and are therefore used widely in PA applications.

Various sensors have been developed for each of these platforms for collecting information about crops or soil. The sensors are divided into two types based on the information collected:

• Imaging sensors record information as aerial photographs or remote-sensing images.

• Non-imaging sensors produce information as numerical data without images.

Sensors can be classified as passive or active, based on their energy use.

• Passive sensors rely on natural light for collecting information. For example, ordinary digital cameras capture images; multi-spectral satellites make use of sunlight to produce images of the Earth's surface.

• Active sensors produce their own light to illuminate and measure the reflected light from the targets. For example, radar sensors can generate thousands of tiny pulses per second; these pulses interact with vegetation or crops and reflect at various levels of intensity.

Four important characteristics of sensors determine their suitability for the different PA applications

• Spatial resolution (also known as pixel resolution) — the ability of the sensor to resolve

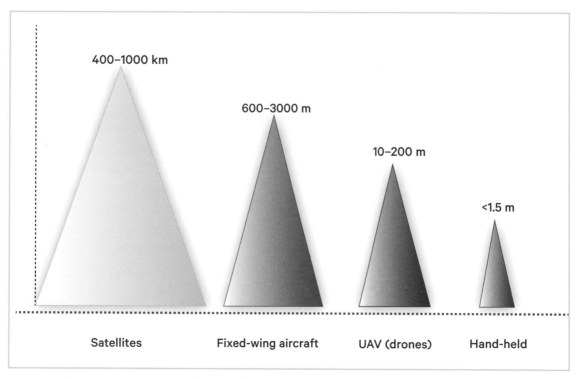

400–1000 km

600–3000 m

10–200 m

<1.5 m

Satellites   Fixed-wing aircraft   UAV (drones)   Hand-held

Figure 12.8 Types of remote sensing platform used in precision agriculture.

details in a photographic image. The higher the resolution, the better the recognition of different objects.

- Spectral resolution — the ability of the sensor to resolve different light (wavelength) intervals.
  - For example, an RGB camera uses three wavelength bands, red, green and blue. A multi-spectral sensor uses two or more bands, whereas a hyper-spectral sensor uses a continuous range of hundreds of bands.
  - Sensors or cameras with high spectral resolution are expensive but have the potential capability to extract detailed information from crops or soils.
- Radiometric resolution — the sensitivity of the sensor to resolve different grey-scale values.
- Temporal resolution — the amount of time needed to revisit and acquire the data for the exact same location.
  - Satellite sensors monitor crops regularly (fortnightly or monthly).
  - Drones can be flown at any time.

Applications for remote-sensing tools

Compared with soil variability, crop properties vary in complex ways because they are influenced by multiple factors such as microclimate, species, farming operations and infestations of weeds, diseases and pests. The tools discussed above are useful for describing complex crop variability. In PA, multi-spectral and hyper-spectral sensors operating on a variety of platforms have been actively used for nutrient, pest, disease and weed management, identifying physiological stress and classifying different crops.

In nutrient management, sensing is typically conducted during critical stages of crop development when the maximum amount of nutrients (such as nitrogen or phosphorus) are being absorbed by the plants. Once nutrient-deficient areas are recognised or mapped by remote-sensing tools, cost-effective fertiliser application strategies such as variable-rate application can be implemented as required. More fertiliser will be applied to deficient areas and less to areas with sufficient nutrients.

In crop or grassland farming, weed infestation typically occurs in patches. In traditional agriculture, herbicide is applied uniformly across a field. Remote-sensing tools help to map areas of weed infestation so that herbicide can be applied selectively to these areas. As weed patches are small, high-resolution sensors are required for mapping them. Because the maps allow herbicide to be accurately applied to infested areas only, significant amounts of herbicide can be saved across the whole farm. Similar approaches can be implemented for disease and pest management (Oerke et al. 2010).

Applications for proximal sensors

Proximal sensors are cheap and easy to use, and so have become an important element in PA services for measuring soil and crop properties (Pullanagari et al. 2012; Viscarra Rossel, McBratney, and Minasny 2010).

Some proximal sensors are designed to work in contact with objects. Types of proximal sensors include:

- Near-infrared sensors
- X-ray sensors
- Gamma ray sensors — capable of producing information relevant to water and potassium content, and bulk density

- Electrochemical sensors
- Electromagnetic (EM) sensors — use electric circuits to measure soil properties such as salinity, moisture and texture at different depths. Some commercially available EM sensors can be towed behind a vehicle (Figure 12.9). The EM map produced shows the variability of soil: areas with higher values have a higher yield potential.
- Mechanical sensors — interact with the soil profile to determine soil strength.

Precision variable-rate equipment

In PA, the information collected is of fundamental value in adopting different treatments that aim to decrease variability within a field. As discussed previously, field variability is identified using various methods, and prescription maps

can then be developed for site-specific treatments. GPS-guided variable-rate technology then plays a crucial role in applying the desired amount of product (fertiliser or pesticide) at each location according to the requirements described in the prescription map. Electronic rate-control systems are designed to adjust the release of different products from the application machines, with different systems having been developed for different products. The main uses of variable-rate technology are:

- fertiliser application
- lime and gypsum application
- agrochemical application
- sowing
- tillage.

For the application of liquids, a flow-based control system is commonly used. Flow systems use pressure valves to control the flow rate over a certain time-frame, releasing the product from the nozzle at the desired flow rate and amount. Multiple nozzles are fixed on a sprayer boom

Figure 12.9 On-the-go EM survey using an ATV.

to cover large widths. These booms are fixed to the rear of a tractor and connected to a large tank. The nozzles are automatically opened and closed as the tractor moves across the field. The tractor driver can control and monitor the process from the tractor cabin.

For the application of dry fertilisers, such as urea, phosphate and lime, a spinning-disc-based controller is mounted on the rear of a truck. The prescription maps are loaded into the truck's computer and the fertiliser is automatically spread based on the physical location of the truck (Figure 12.10). On hill-country farms in New Zealand, the undulating nature of the topography means that fertiliser is generally spread from top-dressing aircraft. Ravensdown recently introduced their Intellispread system which enables variable-rate application of fertiliser using computer-controlled hopper doors.

Precision irrigation

Farming consumes 70 per cent of freshwater globally. As the world's population grows and becomes increasingly urbanised, water resources are becoming limited. Climate change is accelerating the problem through global shifts in precipitation and temperature patterns.

Traditional irrigation systems ignore the variability present in a field and can over-irrigate some areas, resulting in poor growth and nutrient leaching, while under-irrigating other areas which become subjected to water stress. Efficient water management plays a crucial role in establishing sustainability in both current and future farming practices.

Precision irrigation, also known as variable-rate irrigation (VRI), considers the spatial and temporal variability of soil moisture status

Figure 12.10 Variable-rate fertiliser spreading.

and supplies water as per plant requirements. VRI technology enables the application of irrigation water at variable rates in different management zones, using centre-pivot systems. The management zones in this context are homogeneous field areas with a uniform soil moisture distribution. Soil moisture zones are defined by measuring the physical properties of the soil (Hedley et al. 2013).

VRI involves the following steps:

- An EM survey is conducted across the field to determine the soil's water-holding capacity.
- After analysing the data, prescription maps with management zones are constructed and uploaded into an irrigation system.
- In each zone, soil moisture sensors are installed to record soil moisture status. These sensors regularly communicate with the irrigation control system to operate centre-pivot irrigators to maintain optimal moisture conditions.
- Irrigation schedules are prepared to make decisions on how much water to apply and when to apply it.

Benefits of VRI technology

- VRI technology enables water to be supplied according to crop demand, saving 30–40 per cent water overall.
- Supplying optimal water for crop growth also improves crop yield.
- VRI maintains optimal soil moisture conditions and reduces run-off and leaching of nutrients.
- The VRI system is completely automated and can be easily controlled using mobile apps.

Robots

Farmers are increasingly under pressure to feed more people. Farms need skilled workers, and worker performance can be inconsistent for a range of reasons. In contrast, robots can work 24 hours a day in most weather conditions with consistent performance.

Farm robots have been designed to offer services such as drilling, seeding, spraying and harvesting. A start-up by Swiss company ecoRobotrix has developed a fully autonomous weeding robot (Figure 12.11). In a European project named Sweeper, a harvesting robot has been developed for sweet peppers (capsicums) (Figure 12.12). This robot can harvest one fruit in less than 15 seconds.

Precision livestock farming

Livestock farming also benefits from using PA technologies to achieve sustainability. Precision livestock farming (PLF) aims to provide sufficient information to the farmer about the animals to achieve better productivity, health and welfare. PLF is mainly focused on developing livestock management and monitoring systems. It uses sensors and information technology to recognise animals, allowing the system to monitor animal behaviour, characterise individual animals, detect fertility and disease, quantify gas emissions and more. Compared with cropping, livestock farming is more complex and involves ethical issues, so careful management is essential.

Figure 12.11 Autonomous weeding robot by ecoRobotix.

GPS-based monitoring collars

These are electronic units housed in a collar around the neck of individual animals. The collars track animal activity, behaviour and temperature. Some commercial collars also act as a virtual fence (VF), providing digital boundaries without the need for physical fences. The VF allows farmers to improve management strategies by keeping the animals within productive areas and away from environmentally sensitive areas. The collars track animal movement via GPS, and emit audio cues and electrical stimuli relative to the location of a particular virtual fence. Farmers can obtain this information in real time so that effective decisions can be made.

Detection of heat in cows is difficult using conventional techniques. An automated detection system helps to detect heat in cows based on monitoring animal movements. A new, innovative company called Afimilk has developed bovine leg sensors and neck collars for accurate detection of heat through monitoring rumination and eating time (Figure 12.13), enabling farmers to make better decisions regarding the management of individual cows.

Recognition of aggressive behaviour in pigs

In pig farming, aggressive behaviour leads to skin trauma, infection and fatal injuries. A computer-vision-based technology can automatically recognise sick pigs and aggressive behaviour — an aggressive pig's behaviour is significantly different from that of healthy, non-aggressive pigs. Mathematical algorithms can be used to detect behavioural patterns captured by vision technology.

Figure 12.12 Robotic harvester for sweet peppers.

Artificial intelligence for assessment of animal welfare

In recent years, technology has revolutionised livestock management techniques. The behaviour and physiological components of animals can be expressed as facial expressions, changes in posture, vocalisations and odours. These elements can be detected and evaluated for insights into animal welfare. The company Cainthus has developed technology based on computer vision and artificial intelligence that continually monitors cow behaviour with respect to eating, drinking, feed locations and movements (Figure 12.14). Analysis of the information collected translates into animal well-being practices, productivity and performance. Another company, Res Animal Health, combines animal behaviour, genetics and clinical test data to provide better understanding of animals.

Robotic milking machines are well known in the dairy industry. They increase work efficiency by replacing expensive labour, and they maintain hygiene. The company DeLaval has introduced a milking system that automatically recognises each cow when it enters the milking shed. The system cleans udders, identifies teats using a vision system and milks automatically. This system allows a farmer to milk twice or three times a day, with improvements in production and profit. The robot is controlled through a touch screen and is capable of remote operation.

Future farms

To cope with a growing population, PA is vital for achieving sustainable production in modern farming. With the introduction of new technol-

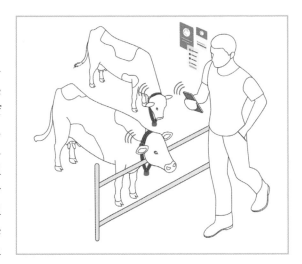

Figure 12.13 Animal neck collars for detecting heat. (Source: afimilk)

Figure 12.14 Computer vision monitoring of cattle.

ogy in agriculture, future farming will be more sophisticated. Technology-enabled farms will be more efficient and sustainable. Currently, most PA technology is in early development only.

In future, robots will become more autonomous and self-sufficient. They will make real-time decisions by analysing big data gathered from multiple sensors. Robots will work on both large and small farms to benefit all farmers. Small groups of autonomous robots will work on different farm operations to deliver precise

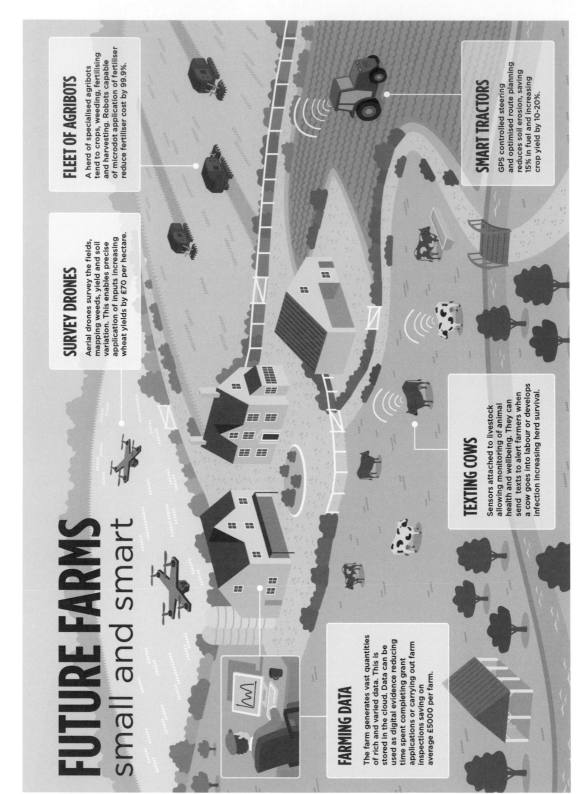

FUTURE FARMS
small and smart

FLEET OF AGRIBOTS

A herd of specialised agribots tend to crops, weeding, fertilising and harvesting. Robots capable of microdot application of fertiliser reduce fertiliser cost by 99.9%.

SMART TRACTORS

GPS controlled steering and optimised route planning reduces soil erosion, saving 15% in fuel and increasing crop yield by 10–20%.

SURVEY DRONES

Aerial drones survey the fields, mapping weeds, yield and soil variation. This enables precise application of inputs increasing wheat yields by £70 per hectare.

TEXTING COWS

Sensors attached to livestock allowing monitoring of animal health and wellbeing. They can send texts to alert farmers when a cow goes into labour or develops infection increasing herd survival.

FARMING DATA

The farm generates vast quantities of rich and varied data. This is stored in the cloud. Data can be used as digital evidence reducing time spent completing grant applications or carrying out farm inspections saving on average £5000 per farm.

Figure 12.15 A conceptual view of future farming. (Source: Nesta)

resources and to reduce waste. As shown in Figure 12.15, future farming will involve the use of advanced PA tools that will be embedded in every stage of the farming process and will be connected to each other to create a smart system and make real-time decisions with minimal human interference. A single person will be able to monitor the process through a digital control dashboard on a smartphone. Such smart farming systems can unlock the secrets of farm interactions and provide precise information about animal health and welfare.

References

Gebbers, R., and Adamchuk, V.I. 2010. 'Precision agriculture and food security.' *Science* 327 (5967): 828–31.

Grisso, R., Alley, M.M., and McClellan, P., (2005). Precision Farming Tools. *Yield Monitor*.

Hedley, C.B., Roudier, P., Yule, I.J., Ekanayake, J., and Bradbury, S. 2013. 'Soil water status and water table depth modelling using electromagnetic surveys for precision irrigation scheduling.' *Geoderma* 199: 22–9.

Oerke, E-C., Gerhards, R., Menz, G., and Sikora, R.A. (eds). 2010. *Precision crop protection: The challenge and use of heterogeneity.* Springer.

Pullanagari, R.R., Yule, I.J., Tuohy, M., Hedley, M.J., Dynes, R.A., and King, W.M. 2012. 'In-field hyperspectral proximal sensing for estimating quality parameters of mixed pasture.' *Precision Agriculture* 13 (3): 351–69.

Shannon, D.K., Clay, D.E., and Kitchen, N.R. 2020. *Precision agriculture basics.* John Wiley & Sons.

Viscarra Rossel, R.A., McBratney, A.B., and Minasny, B. (eds). 2010. *Proximal soil sensing.* Springer.

Whelan, B. and Taylor, J. 2013. *Precision agriculture for grain production systems.* Clayton, Vic., CSIRO Publishing.

Yousefi, M.R., and Razdari, A.M. 2015. Application of GIS and GPS in precision agriculture (a review). *International Journal of Advanced Biological and Biomedical Research* 3 (1): 7–9.

About the Contributors

Lydia Cranston comes from a farming background and has always been interested in agriculture. She is currently a senior lecturer in agricultural production at Massey University and on weekends she works on her own sheep and beef farm. Her main area of research is in the use of alternative forages to create more productive, sustainable and environmentally friendly pastoral production systems.

Julian Gorman is a lecturer at Massey University. Prior to this, he worked at the Research Institute for Environment and Livelihoods, Charles Darwin University, Australia. He has a degree in Botany, a postgraduate qualification in Natural Resource Management and a PhD looking at the agribusiness potential of an Australian native plant. Julian has over 20 years work experience relating to the sustainable utilisation and commercialisation of native plant products.

Kerry Harrington is an associate professor in Weed Science at Massey University, where he has been lecturing students on weed control since 1983. He has also been involved with many research projects over this time on how best to control weeds, both in agriculture and horticulture, including within pastures, arable crops, fruit crops, turf and forestry.

David Horne's research focuses on the impacts of land use on water quality and quantity and the mitigation of adverse effects. This work includes implementing a range of mitigation measures at both the farm and catchment level, such as animal housing, grazing techniques, effluent management, irrigation management, alternative pasture species and cropping techniques. He is also interested in the research of alternative technologies for managing drainage pathways for improved environmental outcomes and water-use efficiency.

Peter Kemp is professor of Pasture Science and was head of the School of Agriculture and Environment, Massey University, from 2009 to 2020. Since 1985 Peter has taught courses in agriculture, and horticulture, undertaken research in pasture science, agroforestry and agricultural systems and has supervised many PhD and Masters students from around the world.

Huub Kerckhoffs (PhD, Wageningen University) is a plant physiologist who has worked with various laboratories at universities or institutes in the Netherlands, United Kingdom, United States, Japan, Australia and New Zealand. He has extensive experience in plant/crop science/horticulture, in particular within vegetable and fruit production systems. He led Massey Horticulture until 2020, and is now with the

Ministry for Primary Industry (MPI), involved in horticultural sector policy.

Marion MacKay is a horticulturalist and environmental manager who specialises in plant biodiversity and conservation. Over many years she has studied both indigenous and introduced plants in New Zealand, examining the diversity of species present along with their management and conservation in botanic gardens and collections.

James Millner's teaching responsibilities include contributing to undergraduate courses in the Bachelor of Agricultural Science and Bachelor of Agribusiness as well as supervising postgraduate students. His research interests include crop production and farm forestry, with current research including the use of native shrubs for sheep browse on hill country and the influence of different forages on nitrate leaching under sheep.

Alan Palmer spent time mapping soils with DSIR Soil Bureau before joining Massey University in 1984. His areas of interest include soil mapping, soil properties, land use, land evaluation and Quaternary geology. At present he has a particular interest in whole farm plans and how they may be used by farmers and advisors to make land use both more profitable and more sustainable.

Reddy Pullanagari received a BSc (Agriculture) and MSc (Agronomy) from an agriculture university in India, and a PhD in precision agriculture from Massey University. He is currently working as a precision agriculture scientist at Massey's AgriFood Digital Lab. His research focuses on developing advanced technologies in agriculture to boost the productivity of the primary industries under safe and sustainable environmental conditions.

Nick Roskruge (Ātiawa, Ngāti Tama-ariki) is an associate professor of horticulture, ethnobotany and Māori resource and environmental management at Massey University. In 2013 he received a Fulbright award and spent time at Cornell University, New York, investigating ethnobotany and potato genomic programmes. Nick is involved in a wide range of Māori and Pacific horticultural projects, and chairs Tāhuri Whenua National Māori Vegetable Growers Collective. His professional activities focus on food security and crop systems in New Zealand, the South Pacific nations and the Americas, particularly the US, Chile and Peru.

Claire Scofield is a research associate at Plant and Food Research. A member of the fruit crops physiology science group, she is involved in research across a diverse range of tree and vine crops, but mainly focused on summerfruit (stonefruit). Her main research focus in the summerfruit area is on new apricot selection evaluation, including tree physiology and postharvest physiology. Other areas of interest are new orchard growing systems, particularly the new narrow-row planar systems, their light environments and resulting fruit quality.

Svetla Sofkova-Bobcheva is a senior lecturer in Production Horticulture in the School of Agriculture and Environment, Massey University. She is also a supervisor for postgraduate and doctoral students in production horticulture, and quantitative genetics and plant breeding. Her research interests focus on plant breeding, in particular developing better breeding strategies in response to abiotic (climate change) and biotic constraints in the context of food security and nutrition. She specialises in horticultural crops and how the genetic potential of a variety can be fully expressed by cultural and agronomical means.

Kevin Stafford is a veterinarian with interests in many aspects of agriculture. His research interests include farm animal production, behaviour and welfare. He has published several books on livestock production and welfare and authored more than 200 refereed papers. He has a small sheep farm and a large garden and orchard.

Jill Stanley is Science Group Leader, Fruit Crops Physiology, for Plant and Food Research in New Zealand. She leads a group of over 40 researchers around New Zealand and in Australia who carry out physiology research on fruit and nut crops to enhance their value and productivity. Her research currently focuses on the physiology of summerfruit (stonefruit) and new apricot evaluation, including the enhancement of fruit quality, regulation of fruit development, production efficiency, tree training, orchard growing systems and postharvest physiology. She also carries out research on berryfruit and the New Zealand native snowberry (*Gaultheria*).

Image Credits

Cainthus: page 261

S. Carrick: page 14 (a, b and c)

Lydia Cranston: pages 35; 39; 42; 44; 45; 46; 47

Kerry Harrington: pages 211; 212; 213; 214; 216; 218; 219; 220; 223; 225

D. Kimber: page 186 (lower)

Marion MacKay: pages 186 (top and middle); 187; 188; 190; 192; 193; 194; 195; 196; 197; 198; 199; 201; 203

McCarthy Contracting Ltd: page 258

James Millner: pages 59; 61; 68; 233; 237; 238; 239

Alan Palmer: pages 14 (d); 15; 19; 21

PGG Wrightson Seeds: page 48

Plant and Food Research: page 123

Pullanagari Rajasheker Reddy: pages 251; 252; 257; 260 (top)

Robotics Plus: page 178

Nick Roskruge: page 91

Shutterstock; pages 2–3; 8–9; 24; 32–33; 49; 54–55; 67; 70–71; 84; 89; 94–95; 107; 109; 112–113; 120; 127; 129; 132–133; 140; 154 (top); 158; 162–163; 169; 176; 179; 182–183; 202; 208–209; 226–227; 242; 246–247; 270–271

Svetla Sofkova-Bobcheva: pages 149; 154 (middle and lower)

Sweeper Consortium: page 260 (lower)

Zespri: pages 171; 174; 180

Acknowledgements

The production of this book would not have been possible without the help of the members of the College of Sciences of Massey University, and in particular Professor Peter Kemp, whose support throughout the process was invaluable.

Agriculture and horticulture underpin New Zealand's economy and are going to be ever more important as we move into a market environment demanding high quality, ethically produced foods. New Zealand is well set to meet these demands but will depend on ongoing scientific progress to do so. The sciences underpinning food production are broad and scientific knowledge grows each year. Thus, there is a need for an easy-to-read introduction to agriculture and horticulture for high school students studying these subjects. This textbook is also suited to university students studying agriculture and horticulture and, also those involved in human nutrition, dietetics, agri-commerce and environmental studies. This primer is also written for the public who have an interest in how their food is produced and for those who write about food in the various media outputs.

Thanks go to all those who wrote chapters for this book for their enthusiasm and hard work and for being willing to produce easy-to-read descriptions of their specific fields of interest.

Indispensable guides for students and practitioners from Massey University Press

Livestock Production in New Zealand

Edited by Kevin Stafford

304pp, $55

This comprehensive book is an indispensable guide to the management of dairy cattle, beef cattle, sheep, deer, goats, pigs, poultry, horses and farm dogs in New Zealand. Written mainly by experts from Massey University's Institute of Veterinary, Animal and Biomedical Sciences, it's of value and interest to everyone from students to farmers, right across New Zealand's agribusiness sector.

The Sheep

Dave West, Neil Bruère and Anne Ridler

408pp, $79.99

This technical and specialist guide to sheep health, disease and production for veterinarians, farmers, farm advisors and veterinary, agricultural and applied science students. Its authors are recognised internationally in the field of sheep health and production and have a lifetime of experience in the New Zealand sheep industry as well as wide exposure to sheep farmers working with practical production and flock health issues.

Diseases of Cattle in Australasia

Edited by Tim Parkinson, Jos Vermunt, Jakob Malmo and Richard Laven

1176pp, $365

The definitive and authoritative text on cattle diseases in New Zealand and Australia. It covers all the important diseases of cattle, with particular emphasis on clinical examination, diseases of the gastrointestinal tract, lameness, mastitis, and reproductive disorders. The chapter on practical therapeutics for the cattle veterinarian is of great value for veterinary students, and during on-farm consultations.

Veterinary Clinical Toxicology

Kathleen Parton, Neil Bruère and Paul Chambers

435pp, $75

A guide to the diagnosis and treatment of poisoning from mineral and inorganic poisons, pesticides, miscellaneous poisons, anthelmintic toxicity, photodynamic agents, mycotoxicosis, corynetoxins, poisonous plants and veterinary drug toxicities.

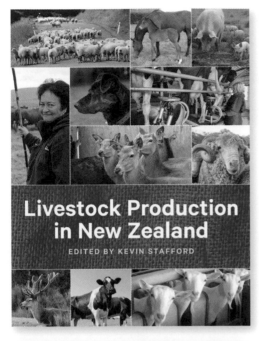

Livestock Production in New Zealand

EDITED BY KEVIN STAFFORD

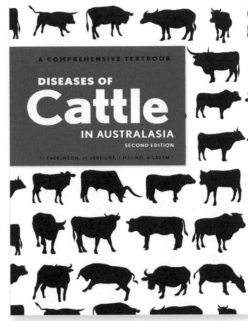

A COMPREHENSIVE TEXTBOOK

DISEASES OF
Cattle
IN AUSTRALASIA
SECOND EDITION

TJ PARKINSON, JJ VERMUNT, J MALMO, R LAVEN

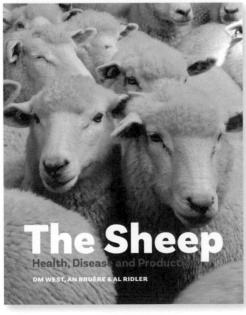

The Sheep
Health, Disease and Production

DM WEST, AN BRUÈRE & AL RIDLER

MASSEY TEXTS

3RD EDITION
Veterinary Clinical
Toxicology
K. PARTON, A. N. BRUÈRE, J. P. CHAMBERS

For more information about our books please visit
www.masseypress.ac.nz

First published in 2021 by Massey University Press
Private Bag 102904, North Shore Mail Centre
Auckland 0745, New Zealand
www.masseypress.ac.nz

Design by Kate Barraclough
Typesetting by Sarah Elworthy
Cover: David Wiltshire (centre front); Shutterstock

A catalogue record for this book is available from the
National Library of New Zealand

Printed and bound in Singapore by Markono Print Media Pte Ltd

ISBN: 978-0-9951230-4-5